PROGRAMMING THE PROPELLER™ WITH SPIN™: A BEGINNER'S GUIDE TO PARALLEL PROCESSING

OTHER BOOKS BY HARPRIT SINGH SANDHU

An Introduction to Robotics
This book, published in 1996, introduces you to robotics and then shows you how to build a weight-shifting, biped, humanoid robot that you can run from a PC. Complete plans are included and all the work can be done with a few tools at the kitchen table. This is the book that laid the foundation for the now abundant humanoid walking-robot industry.

Making PIC Instruments and Controllers
This is a hands-on tutorial and resource book that teaches you how to build your own instruments and controllers using PICBasic on PIC microcontrollers.

This PIC-based book is aimed at connecting your PIC to real-world measurements and sensors. This 316-page book is laid out in a very clear manner that makes it an excellent reference or textbook.

Running Small Motors with PIC Microcontrollers
This book is a hands-on tutorial and that teaches you how to run all sorts of small motors with PICBasic and PIC microcontrollers. A hands-on approach is brought to this PIC-based book aimed at running small motors of all sorts with these microcontrollers. Over 2,000 lines of PICBasic code you can use are included. This 352-page book is laid out in a very clear manner that makes it an excellent reference or textbook.

Spindles
This book is a treatise on working with the small metal-cutting lathe. It concentrates on making spindles that allow you to do simple milling machine work in the lathe, such as cutting gears and making clockworks.

Convenience Items
Certain items that make experimentation with the Propeller easier are available from Encodergeek.com, which is the support site for this book. Go to this site for book support, more illustrations, prices, and descriptions.

PROGRAMMING THE PROPELLER™ WITH SPIN™: A BEGINNER'S GUIDE TO PARALLEL PROCESSING

Harprit Singh Sandhu

New York Chicago San Francisco Lisbon London Madrid
Mexico City Milan New Delhi San Juan Seoul
Singapore Sydney Toronto

The McGraw-Hill Companies

Cataloging-in-Publication Data is on file with the Library of Congress

McGraw-Hill books are available at special quantity discounts to use as premiums and sales promotions, or for use in corporate training programs. To contact a representative, please e-mail us at bulksales@mcgraw-hill.com.

Programming the Propeller™ with Spin™: A Beginner's Guide to Parallel Processing

1 2 3 4 5 6 7 8 9 0 WFR WFR 1 0 9 8 7 6 5 4 3 2 1 0

ISBN 978-0-07-171666-6
MHID 0-07-171666-1

Sponsoring Editor
Roger Stewart

Editorial Supervisor
Patty Mon

Project Manager
Vastavikta Sharma, Glyph International

Acquisitions Coordinator
Joya Anthony

Copy Editor
Bart Reed

Proofreader
Claire Splan

Indexer
Jack Lewis

Production Supervisor
George Anderson

Composition
Glyph International

Illustration
Glyph International

Art Director, Cover
Jeff Weeks

Cover Designer
Jeff Weeks

This effort is dedicated to

Robert A. Hoffswell, BA, MA
Mathematician, Engineer, Scientist, Gentleman

"My first acquaintance from the Age of Information"

"Harvey the Fox" is a mathematician who worked as a physicist at the University of Illinois Particle Accelerator Lab. In the very lab where the cyclotron was invented. Where a copy of the original model hung on the wall.

He is my first acquaintance from the Information Age. A man who knows how to research and understand things he once knew nothing about and how to use the information to create exquisite, useful, and fully functional devices.

He is an Engineer and Scientist of course, but he is also a ham radio operator, a journeyman carpenter, a master woodworker, a maker of musical instruments, like fine guitars, ukuleles, and harpsichords. And exquisite looms and other wonders along the way.

I have learned much about learning from the Fox.

CONTENTS

PREFACE

After I finished my book *Running Small Motors with PIC Microcontrollers*, I asked my friend David H. at HVW Technologies in Canada if he had in any ideas as to what might be worth covering in my next book. David suggested that a book about the new Propeller chip from Parallax, written in the same vein as my other hands-on books, could be a welcome effort. With this in mind, I contacted Parallax, Inc., in California and they turned me over to Ms. Stephanie Lindsay, their contact person for authors. Ms. Lindsay was good enough to send me a comprehensive authoring package to get me started on this adventure. In this book I share what I have learned about the Propeller chip and parallel processing with you. It is my wish that by the time you have read through it and have done all the experiments, you will have the confidence, skills, and knowledge necessary to start using the Propeller chip in ways that will make your life both more interesting and, hopefully, more productive.

My first reaction to opening the authors' package and starting on the Propeller manual was, How am I ever going to learn to use this processor? The material was not beginner friendly. Although it was at a higher level, it was very interesting. The further I got into reading and understanding the manual, the more fascinated I became with what the very clever engineers at Parallax had created. It is certainly one of the wonders of the modern world that you can buy eight 32-bit processors and shared memory for less than $8. In this book we will discover what all this, including parallel processing, means to us as engineers, technicians, and hobbyists. As always, I will minimize the use of complicated formulas and jargon so that if you are interested in things mechanical and electronic and have a rudimentary knowledge of what a computer program is, you will be able to use these processors to undertake simple tasks and maybe even some fairly complicated projects in a parallel-processing environment.

There are, of course, two aspects to learning how to use the Propeller chip. The first is learning how to use each of the identical 32-bit processors in the chip. Parallax calls each of these eight processors a "cog," and each of these cogs is similar to a typical 32-bit processor, with some special features added and some left off. The second is learning how to make these eight cogs interact with one another in an effective way to explore the fascinating parallel-processing possibilities that are now suddenly within our reach. Because the eight 32-bit processors on the chip are identical, once you have learned to use one of them, you have learned to use all of them.

The intellectual discipline that has to be mastered involves setting up the problem in such a way that the eight processors can be used in the most effective way possible, thus creating a viable solution for the task you have in mind. This has to do with really understanding the problem and with learning how to break a problem down into separate tasks, each of which can be assigned to one of the cogs, in an orderly and logical way. We will learn how to do this.

Because not every problem lends itself to a parallel-processing solution, we will spend some time on learning how to identify those problems that can be solved within a parallel-processing environment.

This book is intended for the novice user. It is for the novice user for two reasons: One, the material that Parallax provides regarding this chip is more advanced than a first-time user can master with ease. Two, I am also a novice as far as this particular discipline (parallel processing) goes. In this book, I share what I have learned with you, in a straightforward and hopefully nonintimidating manner that you will find useful. We will learn by doing, which is the best way to learn to do anything. Before we can start, though, we have to understand what the system is capable of and how this book is organized to address the tasks at hand.

This book is divided into four parts that compartmentalize what we are interested in:

- The first part of the book introduces you to the Propeller chip, starting with one cog (the term used by Parallax for each of the eight 32-bit processors in the Propeller chip). First, we learn about just one of the eight processors and how to interact with the I/O provided on the chip. This I/O is in a shared portion of the chip that all the cogs can address. All the features of the one cog are covered in detail. At the end of the first part, you should have a good idea of what the system is all about and be ready to begin using it.
- In the second part of the book, I cover what you need to know to develop the skills necessary for creating a system that allows the cogs to work together. We will do this by learning how to make the Propeller interact with a number of input and output devices. I have selected the kind of devices an amateur enthusiast, a technician, or an engineer is likely to be interested in interacting with on a day-to-day basis—displays, switches, detectors, motors, and such. The limited memory in the system does not lend itself to the handling of large arrays and related number-crunching applications, so simple control applications like the ones we will consider are the most suited for investigation by us beginners. The major objective of this part of the book is to learn how to read and create signals of various types.
- In the third part of the book, we use the lessons we have learned in Part II to build and program a number of devices using the Propeller chip. The device in each experiment is a real-world application of the parallel-processing environment, and each one uses more than one cog. Not all the projects are completed 100 percent, so you have the challenge to complete them. Appropriate information and hints are provided.
- The fourth part of the book is composed of appendixes that provide you with supplemental information you will find helpful in using the device. This includes the hardware and software resources needed for the experiments we undertake. Where special items are needed, I have made arrangements to provide them on my website at Encodergeek.com.

A large part of the information you need to help you use this processor can be found on websites maintained by Parallax and others. I recommend that you get familiar with what is in these websites and get comfortable with using the material these websites provide. The discussion forums are extraordinarily useful and should be made a part of your regular reading and learning experience. The following three online resources provide a good starting point as you progress through this book:

- The Parallax forums. These forums are your most useful resource.
- Wikipedia provides useful general information.
- The support web pages provide specific information.

The Propeller chip software is organized such that routines called methods and procedures, created by one person, can be used by others with relative ease. Most of these procedures are well documented, and because all the code is visible, you can modify it to serve your more specific needs should that become necessary. This being the case, becoming familiar with and studying the work done by others is one of the skills you need to develop. I will provide methods with full documentation to support the devices we use so that you can see how these methods are created and then called in subsequent programs.

All the experiments in this book can be undertaken with the Propeller Education Kit (32305) provided by Parallax and a minimal amount of additional hardware. Other than the Propeller programming tools provided by Parallax at no charge, consisting of Spin (a high-level language) and PASM (the Propeller Assembly language), no other software is needed. The few extra hardware items needed are listed in Appendix B. The work that we will undertake will be designed around the creation of software for running the type of devices you might use in the design of day-to-day projects on the hobby bench, the engineering laboratory, or the industrial research facility. The devices selected are inexpensive and fit within the constraint of working well with beginners who are just learning how to use the Propeller chip. None are hard to use.

As mentioned previously, Parallax provides two languages for programming the Propeller chip. The first language is a high-level language called Spin. The Spin language can be used for almost everything we need to do, except for tasks that need extreme speed and therefore have to be handled with some sort of Assembly language constructs. All the work in this book will be done in the Spin language, with minor calls to Assembly language routines if and when necessary. The goal is to become familiar with the Spin language and to be aware of the capabilities that Assembly language provides. Assembly language routines can be embedded in the Spin language programs with minimal effort. In that this book addresses the needs of beginners, we will not do any programming in the Assembly language PASM.

A large part of this book is dedicated to getting an understanding of how the Spin language is used to manage an eight-processor system and its shared memory. A number of simple rules are formulated to allow you to do this from a beginner's point of view. Starting and stopping cogs and assigning specific tasks to them is covered to give you a feel for how you might use these functions. Because there are only eight cogs on the chip, we have to have some discipline regarding how they are to be used. Because each cog is also capable of performing more than one task, as is any "run of the mill" processor, we also need some understanding of when and why more than one task should be assigned to one processor. This, too, is discussed so that you can assign tasks in a more logical manner when you design the hardware and software for your projects.

The Spin language does not support an interrupt capability of any description. Having eight processors in a parallel configuration pretty much eliminates the need for interrupts. However, there are still times when you may need to assign some form of more immediate attention to a task, and techniques that can be used to achieve this are demonstrated.

Attention is also given to the special features each cog supports. Of advanced interest is the use of the two counters provided in each cog and the interesting ways in which they can be used. The use of these counters is not obvious to the first-time user, especially so if he is a beginner. These counters provide an important and powerful function within each cog, and using them effectively is an important part of using the Propeller chip. Because there is a total of 16 of these counters in the Propeller, they provide a resource we cannot ignore if we are to consider ourselves as being familiar with the Propeller system. Parallax provides detailed information on using these counters, but most of it is beyond the understanding of beginners. We will use a counter to create a PWM signal when we need one; no other use is covered.

Part III of the book is devoted to the projects. I have concentrated on controlling the types of things beginners will be interested in. I have covered some of these tasks in other books I have written, and other authors have written a lot about these topics as well. The difference in this resource is that we concentrate on how to undertake these tasks with the Propeller chip and its parallel-processing environment in a way that's interesting to the beginning programmer. Controlling things also has a lot do with process control, so in a way this book is a simplified introduction to process control in the parallel environment.

The one thing that the Propeller chip does not do well is handle large amounts of data simultaneously. Such data crunching requires three basic resources:

- A fast processor to perform the work quickly. (This we have.)
- The implementation of a standardized, verified math package in the software to allow for sophisticated mathematical manipulations.
- A large amount of memory to store all the data to be crunched.

The Propeller chip does not support all three of these capabilities and implementations. This does not mean that we cannot do the day-to-day math we need for our control operations; however, it does means that we are not working with a machine designed for crunching large number arrays. Consequently, there is no discussion of problems that require sophisticated mathematical capabilities in this resource.

Then again, these are not problems that the average beginner is expected to need to address.

I have permitted myself a certain amount of repetition from time to time in the various chapters to allow each chapter and each program to stand alone so that you do not have to read the entire book or look back for segments of programs to get the information you need from any one location. Almost all program listings provide complete programs that are ready to run.

In this book I use the word "transparent" to mean invisible to us. Something we see right through without knowing it is there. Those aspects of a program's operation that are invisible are described as being "transparent to the user." To the beginner, this means that those things that are transparent do not need to be of concern at this stage of the learning process. They happen automatically in the background, and the beginner cannot see or manipulate any aspect of their operation. We can ignore them for now. An example of a transparent operation is the operation of the computer mouse. It does its work without ever making any aspect of its internal workings visible to the user.

The information in this resource came from all sorts of sources I was exposed to in my research, and they are not important enough to be documented as footnotes. The most important of them are the Propeller Manual, the Parallax forums, the Propeller object exchange, and the Internet. The experiments and exercises are similar to those I have used in my books on the PIC 16F877A about making instruments and controllers and running motors. These are basic techniques I developed to explain how the various techniques are implemented within a microprocessor. The task we are undertaking does not change, but the techniques used to get the job done do so that we can accommodate the instruction set available for the logic engine being used.

The basic hardware you need to get going is the Propeller Education Kit (32305) from Parallax. Arrangements have been made to allow experimenters to get all other hard-to-get items from my Encodergeek.com website, and a list of what you need is provided in Appendix B.

The Encodergeek.com site also hosts the support information and updates for the contents of this book. All the programs in the book are provided on the Encodergeek .com website, and you can copy and run them from there if you like. The site also contains a lot of other information of interest to the amateur experimenter.

All the software in this book is provided for your use under the terms of the MIT License. It is yours to use as you see fit. Here is a commonly used version of the statement of the license.

Terms of Use: MIT License

—Harprit Singh Sandhu
Champaign, Illinois USA

harprit.sandhu@gmail.com

Part I

THE PROPELLER/SPIN SYSTEM

Introduction for the Beginner

Before we can start doing things with the Propeller, we need to have an understanding of exactly what we have to work with. Once we understand the Propeller hardware and a bit about the Spin software, we can proceed with our experiments and develop the software needed to execute them.

1

A GENERAL INTRODUCTION TO THE PROPELLER CHIP

Notes *A data sheet is now available for the Propeller chip, under the Help section of the Propeller Tool (Version 1.2.6). If you need specific information about the Propeller, you should refer to the data sheet. The very adequate Propeller Manual (Version 1.1) is suitable for beginners and can be downloaded from the Internet at no charge. Teaching materials are being developed and released by Parallax on a continuing basis. Discussion forums are in place and provide very helpful advice. An extensive Object Exchange is maintained by Parallax on their website. However, all current efforts are aimed at those already fairly comfortable with microprocessor programming. This book provides an information resource for the beginner. It attempts to fill the void with a simple "learn by doing" approach designed specifically for the beginner who knows very little about this chip but has some general familiarity with electronics and with microprocessor programming.*

Before we start I would like to share a secret with you: Once you learn to use the Propeller in its parallel processing environment, you will be hard pressed to ever again use a conventional processor for the kind of tasks small microcontrollers are designed for. The Propeller system is both incredibly powerful and incredibly easy to use. You will be glad to have learned how to use it.

The Propeller Manual

If you need to know more about any of the topics discussed in this book, go to the Propeller Manual for details. The information provided in the manual is not repeated here.

This being the case, I strongly recommend you obtain a copy of the Propeller Manual to use as your absolute reference as you study these chapters. A copy of the manual is also available in the Help section of the Propeller Tool. You should have the electronic copy open on your computer when programming the Propeller so that you can perform an electronic search when you need to find something in the manual. You will find that having both a hard copy and an electronic version of the manual is convenient and useful.

Before we can start using any device effectively, we need to have some familiarity with the general framework within which the device operates. In our case we need to gain an understanding of how the eight microprocessors (the cogs) on the chip are arranged to interact with the shared memory and the other system resources that tie the eight cogs together. In Part I of the book in general, and in this chapter in particular, we will gain an understanding of these arrangements as we discuss some other general aspects of parallel computing in an introductory format.

In 2006, some very clever and gifted engineers at Parallax, Inc., in California, under the leadership of Chip Gracey, introduced "Parallel processing for the rest of us" to the world. This adventure comes to us in the form of a chip that contains eight 32-bit processors with a rather large amount of shared memory. (It is available in a number of formats, including the 40-pin DIP package we are using.) This chip is called the Propeller. Along with the hardware in the form of the chip, the team has provided us with software in the form of two languages suitable for programming the chip. The first language is an object-based language called Spin, with some interesting and useful formatting enhancements. The other language is called Propeller Assembly and consists of a set of assembly language routines that allow fast, more immediate and elemental programming of the Propeller chip. All higher-level programming is done in the Spin language, and all assembly language references can be executed within the Spin framework so that an assembly language program can be started and stopped with just a couple lines of Spin code bracketing the entire assembly language notation. Programs are written with the Propeller Tool (an editing and loading program provided by Parallax). All the software is free for the downloading, from the Internet!

All this leaves all the other systems you might consider for the kind of work that microcontrollers do in the dust. So much so that during the Fall of 2009, even Parallax did not fully comprehend what a wonderful logic engine they had created, and for a couple months you could not buy a 40-pin DIP Propeller chip anywhere in the world. Everyone was out of inventory. (The LQFP and QFN versions of the chip remained available.)

Together, the hardware and software environments provide us with everything we need to implement small parallel processing scenarios. Because parallel processing is pretty much accepted as the next big step that will have to be undertaken to speed up the computational processes currently being run by fairly fast single-processor linear systems, the ability to play with these eight processors in a well-managed environment is a dream come true for many of us. In this book, I will share this adventure with you and in the process of so doing will introduce you to the techniques and rules that allow you to use the Propeller system with confidence and maybe even some expertise.

Parallax has published a detailed data sheet for this chip. It contains all the information required for the kind of processing that needs to be informed of the most intimate details of the insides of the engine. This is not an important resource for beginners, though. The Propeller Manual provided for the Propeller chip covers both languages in detail with more than adequate examples of instruction usage and notation.

Application notes that cover more specialized aspects of the chip's operation are posted on the Parallax website for downloading at no charge.

The discussion forums maintained by Parallax provide additional support that gives you access to everything you need at your level of expertise and understanding. The discussion forums are active, and a number of informed individuals both from within the Parallax organization and outside it post regularly. The range of what people are doing with the Propeller is extensive and impressive, even amazing. I urge you to read these forums regularly and participate in them as often as the need arises.

Parallax also provides "live person on the phone" technical support for the Propeller chip at no charge!

Parallax, Inc.

Parallax, Inc., is located in Rocklin, California. It is a relatively small corporation that provides microcontrollers, development tools, sensors, and robotics for industrial and educational organizations as well as for hobbyists, with their BASIC Stamp®, Propeller™, and other related product lines. Parallax can be reached through the mail at the following address:

Parallax, Inc.
599 Menlo Drive
Rocklin, California 95765
USA

Phone numbers are

Office:	(916) 624-8333
Toll-free sales:	(888) 512-1024
Toll-free tech support:	(888) 997-8267

Their website is:

http://www.Parallax.com/

The discussion forum for the Propeller chip can be found at:

http://forums.Parallax.com

And some other resources of interest include:

Wikipedia
Various independent Internet forums dedicated to Propeller usage

Overall System Description

In the most general of terms, the Propeller chip consists of a bank of eight 32-bit RISC-like (but not true RISC) microprocessors (called "cogs") with some shared memory that also share 32 I/O pins between them. Each of the eight processors can access these 32 I/O pins at all times. The 32 pins are accessed by setting the direction of each pin as either an input or an output (and, of course, the condition of the pin as high or low if it is designated to be an output). In order to coordinate the operation of the eight microprocessors, a hardware device called a "hub" accesses each processor in a round-robin fashion as controlled by the system counter and the system clock. Each cog has access to the system for the same amount of time. The eight cogs are independent of one another and are *identical in every detail.*

The chip provides a bank of memory that can be accessed by each of the cogs when it is the cog's turn to do so, as determined by the hub. Other ancillary functions such as reset delays and oscillator inputs are also provided. In general, the resources available on the chip can be divided into two families, referred to as the common and the mutually exclusive resources. The common resources, which encompass all the I/O pins and the system counter, are available to all the cogs at all times. This means that any number of cogs can access these simultaneously. The mutually exclusive resources, on the other hand, can only be accessed when it is a cog's turn to have control of the system. As mentioned previously, the controlling sequence is managed by the processor hub, which gives access to each cog in a round-robin fashion. Doing this in an orderly way allows the system to stay in sync at all times.

The system clock is a "programmable speed" device that can be controlled by an internal circuit, a phase-locked loop, *and* a crystal oscillator, as may be decided on by the system programmer/designer. The system clock itself is a 32-bit counter that is incremented during each cycle of the system oscillator. The system does not keep track of how many times this counter overflows, and the major use of this counter is for timing delays and other time-related functions that do not need to know how often the clock has overflowed. If you need to know how many times the clock counter has overflowed, you have to design a routine that will do that for you as a part of your program. The operation of the system clock is under the control of the clock register. The clock rate is programmable from 20 KHz to 80 MHz. The hub and the bus operate at half the rate of the system clock. (The slower the clock rate and the fewer the cogs in operation, the lower the power the system uses—and the savings can be substantial.)

Each of the *eight identical cogs* has the following resources within it:

- A 32-bit reduced instruction set (RISC-like) processor
- 2KB of cog RAM
- Two input/output assistants with phase lock loops
- Two powerful counter modules
- A video generator

A number of special-purpose registers are designated in the RAM to read the system counter, to manage I/O pin direction and states, and to configure the counter modules and video generator hardware. The lock bits are special and are located in their own register in the hub; they are accessible from the lock commands only.

The hub allows access to each of the cogs in a round-robin fashion at a clock rate that is half the rate of the system clock. Detailed descriptions of the best- and worst-case scenarios for the interaction of the hub with the cogs, along with timing diagrams, are provided in the Propeller Manual.

All 32 I/O pins can be used for I/O when the system is not using pins 28 to 31 for serial interfacing and/or external memory access. Each of the 32 pins can be made into either an input or output. If made into an output, each of the pins can be set either high or low. When designated as an input, the pin assumes a high impedance state and awaits input from the outside. Because each of the cogs can access all the I/O pins at all times, there is a need for an I/O sharing protocol that keeps things under control. You should become familiar with this protocol before designing complicated schemes for accessing the I/O pins with more than one cog. For our immediate purposes, however, this can be ignored.

A family of eight lock bits is provided by the system to prevent errors when more than one cog is accessing common resources. Conditions can exist such that one cog is writing to a memory location when another cog is reading from that same location. The system of locks enables you to lock down and release the system while you undertake certain critical read/write operations (on more than one 32-bit long). The chip has 64KB total of shared (main) memory that can be accessed by all the cogs on a mutually exclusive basis, as controlled by the hub. Of this, 32KB is in the form of RAM and 32KB is in the form of ROM.

The main RAM is used by the Propeller to store the application we write, along with data, variables, and stack space to perform specified operations.

The other half of the main memory, the ROM, contains the font character definitions and the sine and log tables used for writing to display devices and for creating the math functions needed by any application. Main ROM also contains the boot loader, which is used upon startup. The Spin interpreter is also stored here, and it gets copied to each cog RAM, from where it fetches, interprets, and executes Spin code tokens from main RAM.

The Propeller Tool

The Propeller Tool is a software-writing environment suitable for creating programs in Spin and Propeller Assembly. It is designed for use on an IBM (or Windows/Intel) compatible personal computer. (The software is not available for the Apple Macintosh at this time.) The Propeller Tool is connected to the Propeller hardware environment most conveniently through a prop plug, an inexpensive USB–to–serial port converter available from Parallax. (If you like, you can make your own converter; a circuit is provided in Propeller Help.) Downloading the programs to the Propeller chip and then executing them is a one-keypress proposition.

Parallax provides a special font (called the Parallax font) for writing programs for the Propeller chip. This font is resident in the chip's main ROM and is to be used in video displays. It is also included in the Propeller Tool as a TrueType font that is installed on the programmer's PC and used by the Propeller Tool.

The font is divided into two parts: one for writing programs and the other for creating the wiring diagrams needed to document the effort being undertaken. Special features of the software environment allow extensive documentation to be embedded in the program being written at two levels. Each level of the documentation can be suppressed and made visible as needed. Because this can be done at two levels, minimal documentation can be separated from extensive documentation, and either one or both can be suppressed or made visible as needed by a programmer or user. There is no particular point in going over the details of the operation of the Propeller Tool. You should refer to the Propeller Tool's Help section for these details. At the time of this writing, the current version of the Propeller Tool is version 1.2.6.

The Propeller Tool environment is a very powerful and easy-to-use environment that allows you to have any number of programs open at any one time, limited only by the memory you have available in your personal computer. Having multiple programs open at the same time allows cutting and pasting between programs and speeds up the writing and documentation of programs. Among other things, it allows you to jump back and forth between programs with one click and execute the program with another. This, in turn, allows you to make modifications to programs and run either version to see the changes painlessly. Very ingenious, very powerful.

Instruments Needed to Support Your Experiments

A good volt-ohm meter (VOM) and an oscilloscope should be available on the workbench at all times.

A display instrument we will be using in almost every experiment is the 16-character-by-2-line LCD, but I am not listing it here as a necessity because we are using it as an electronic component onboard the breadboard, as opposed to using it as separate instrument.

Having an oscilloscope is really a must, and learning how to use it effectively is a never-ending life-long learning experience. If you don't have one, get one on eBay as soon as possible. An inexpensive dual-trace 20 MHz scope will be more than adequate for all your needs. An oscilloscope gives you the ability to see things, which is both very powerful and very useful for all investigators.

Note *A 20 MHz oscilloscope is just fine for much higher frequencies. The problem with looking at higher frequencies is that they will not be seen as crisp square waves because the scope cannot respond rapidly enough. You will still see the square waves; it's just that the corners will be rounded a bit. This does not bother what we are interested in. Even a cheap 5 MHz scope is fine. The important thing is to have a scope, and one with dual traces is the scope of choice. This way, you can compare waveforms and see relative timings.*

2

THE PROPELLER CHIP: AN OVERALL DESCRIPTION

The official identification number of the Propeller chip is P8X32A-D40 for the 40-pin version. The Propeller chip is manufactured by Parallax and is available to the general public in single unit quantities, via the Internet, for $8 each as of this writing. For that small amount of money you get eight 32-bit processors that access a rather adequate 32KB shared RAM and 32 lines of I/O. This is unprecedented power and value for those of us who need to use microprocessors to do our day-to-day work in the engineering office, the university laboratory, the technician's workbench, or the hobbyist's workshop. Of course, manufacturers also have a serious interest in this powerful chip, but here we are talking about an introduction for the beginner, so we will not go into what we might do in an industrial environment. Once you understand the basics, moving up to more advanced techniques can be undertaken without difficulty.

Figure 2-1 provides the pin designations for the 40-pin DIP version of the Propeller chip. The Spin language uses the inner P designations for accessing the I/O pins. The external pin numbers are not used. (We will not use them either.)

In this book, all references to the Propeller Manual are made to the latest version of the manual as revealed by Parallax (Version 1.1). The latest version is the version available under the Help menu in the Propeller Tool and on the Internet, where it can be downloaded at no charge.

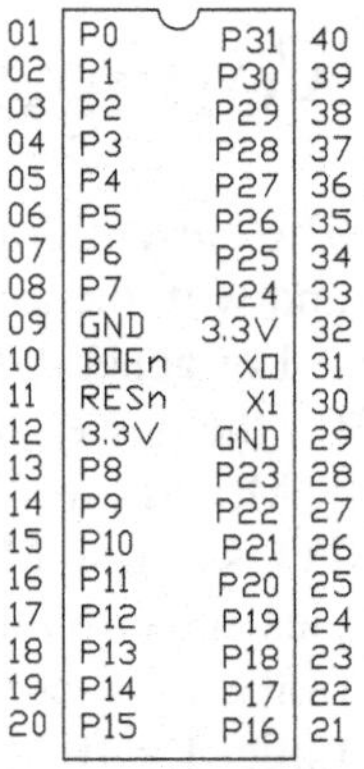

Pin	Name	Name	Pin
01	P0	P31	40
02	P1	P30	39
03	P2	P29	38
04	P3	P28	37
05	P4	P27	36
06	P5	P26	35
07	P6	P25	34
08	P7	P24	33
09	GND	3.3V	32
10	BOEn	XO	31
11	RESn	XI	30
12	3.3V	GND	29
13	P8	P23	28
14	P9	P22	27
15	P10	P21	26
16	P11	P20	25
17	P12	P19	24
18	P13	P18	23
19	P14	P17	22
20	P15	P16	21

Figure 2-1 The pin designations for the 40-pin Propeller chip

Basic Propeller Specifications

Parallax uses bytes (8 bits), words (16 bits), and longs (32 bits) for its memory descriptions. The Propeller chip has the following basic specifications:

- The Parallax model designation for the chip is P8X32A-D40.
- It is a 40-pin DIP chip.
- It runs at 3.3 volts DC (VDC).
- The chip can be run at from DC to 80 MHz.
- It contains eight 32-bit processors called "cogs."
- Each of the eight cogs has 512 32-bit longs as its own RAM.
- There are 32KB of RAM and 32KB of ROM accessible to each cog in a round-robin fashion. (This is hub memory.)
- It has 32 I/O lines that can be addressed by all the cogs at all times.
- Each individual I/O line can sink 32 milliamps.
- Any eight I/O lines can together source a maximum of 100 milliamps at any one time.
- A management scheme allows the eight cogs to access all other resources in a round-robin fashion.
- Any number of the eight cogs can be on or off at any one time. One cog must remain on to allow the system to stay alive.
- Cogs can be loaded with new software in real time.
- Any cog can turn any other cog off and start the next available cog. (They are identical.)

Voltage and Amperage Requirements

The current draw of the device varies with the temperature, the operating frequency, and with the number of cogs active at any one time.

For general design purposes, a requirement of half an amp at 3.3 VDC can be used for the power supply.

More power is used if more cogs are active, and more power is used if more of the I/O is in use. More power is necessary if the processor is being run at higher frequencies because the higher frequencies needed more power to maintain operations. Most of the work done in this book will be done with a 5 MHz external crystal and 2× multiplier for 10 MHz operation.

The Operation of the Eight Cogs

The eight cogs in the processor run independently and simultaneously, and are controlled and coordinated by a system counter or clock. From one to eight cogs can be active

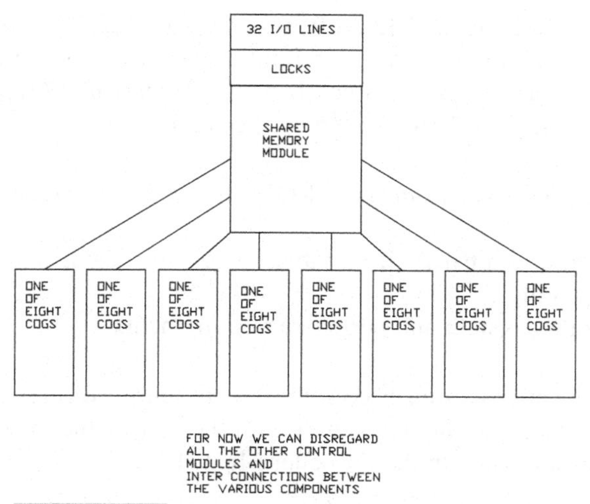

Figure 2-2 The basic layout of the Propeller system

at any one time. All the cogs have access to all the I/O pins, the system counter, the data bus, and address busses at all times. The various activities within each of the cogs in the system are accessed in an orderly, round-robin fashion as controlled by the hub.

The hub does not care how many cogs are running at any one time. It accesses them one at a time in a round-robin fashion. Each hub is accessed for the same amount of time to keep the entire system synchronized with the system counter. When the hub gives a turn to a cog, that cog has exclusive access to the shared resources. The shared resources can be divided into two types of resources: the common resources and the mutually exclusive resources. Each cog has access to the common resources at all times, as mentioned previously, but only one cog, the active cog, has access to the mutually exclusive resources and that's only when it is its turn to be in charge of the system. All this is transparent to beginners, and we do not need to worry about it right now. (When I use the word "transparent," I mean that we are not aware of its operation. In other words, we see right through it and are not aware of its presence, like a pane of glass.)

The part of the system of immediate interest to us is shown in Figure 2-2. Note that this is a very simplified diagrammatic representation. Refer to the Propeller Manual.

The Cogs

As was mentioned previously, there are a total of eight cogs. They are *identical in every way* except for their identification (as Cog_0 to Cog_7). Cog_0 is the controlling cog, on startup, and as you might expect, at least one cog has to be active at all times. These notes provide you with an introduction to the cogs in a simplified format.

Each of the eight cogs is a fully fledged 32-bit RISC-like processor. The usual interrupt structure and its related functions are not implemented in any form in the Propeller system. They are not needed. Each cog has an independent 2KB of RAM, accessible as 512 32-bit longs. This RAM is used by the cog to store:

- The Spin Interpreter, copied over from main ROM, if the cog is executing Spin code
- Program code copied from main RAM, variables, flags, and data, if the cog is executing Propeller Assembly
- Interface locations for other system and peripheral requirements

Certain specific RAM locations in each cog, the special-purpose registers, have been assigned to specific uses to allow interaction with the rest of the system in an orderly manner. These uses are listed in the Propeller Manual.

The Hub

The hub is a hardware device that controls how and when each of the cogs will interact with the various parts of the system. It also controls all the resources needed to maintain the integrity of the system. Because certain timing sequences have to be maintained to allow the system to operate properly, these need to be understood by the programmer so that he or she will design programs that comply with these timing and access requirements. For beginners, there are no special requirements that need to be met. Most of the requirements have to do with critical timing, which will have to be addressed when programming in Assembly language and dealing with timing critical tasks. The timing diagrams for these interactions are illustrated and explained in the Propeller Manual.

Forty Pins Total, 32 Pins I/O

In this book, we are considering the 40-pin DIP version of the chip. Other, much smaller form factors are also available.

Of the 40 pins on the Propeller DIP, 32 are I/O pins. The remaining 8 pins are used for power connections, reset connections, grounding, crystal connection, enabling the system, and other usual and basic housekeeping tasks seen on all microprocessors. All 32 I/O pins can be used for I/O when pins 31 and 32, which are reserved for serial communications to a PC, are not being used for communication. This communication takes place exclusively through pins 31 and 30. Pins 29 and 28 are used for access of external memory when such memory is in use. At other times they can be used for regular I/O connections. Refer back to Figure 2-1 for a pinout diagram of the 40-pin version of the chip.

Connecting to the Propeller

When we are connecting to a low-voltage, low-power device such as the Propeller, our goal is to minimize the load on the pins and to thus protect the device from harmful voltages and currents that might be generated by our experimental circuitry (and our attendant unintentional mistakes).

The Propeller chip is designed for extremely low-power applications and runs at 3.3 VDC. Though this voltage is adequate for interaction with transistor-transistor logic (TTL) devices for most applications, it is desirable to put the output signals through standard (7404 or 7414) gates to keep the current requirements at the Propeller to a minimum. Connections through gates will isolate the chip and limit the power back-fed into the Propeller if we make a wiring mistake. When we use these gates in our designs, they will most probably be inverting gates, meaning that each gate will invert the signal coming into it. We will have to keep this in mind as we write our programs. The advantage of using these gates/buffers is the increased power they provide for driving any attached loads and the isolation/protection of the Propeller chip. Needless to say, a separate 5-VDC power supply has to be provided for these gates and the circuitry beyond them. The ground connection is common.

Noninverting gates (7407s) are available if you prefer to use them, but you will find that for most purposes an inverting arrangement is more convenient. Putting a signal through two inverting gates leaves it unchanged in polarity and is a convenient way of getting a signal buffered but unchanged if you need to do so. Figure 2-3 illustrates the use of an inverting buffer to control an LED.

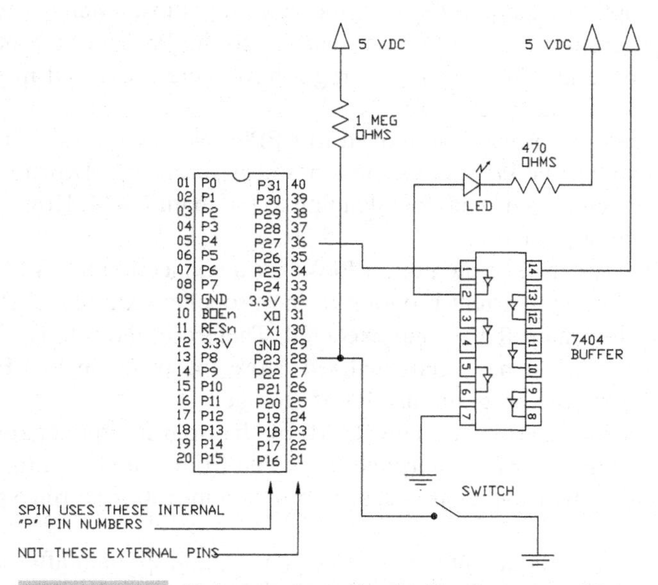

Figure 2-3 Using an inverting buffer to connect to an LED and a dry contact switch with a pull-up resistor

The System Counter

The 32-bit system counter operates at the same rate as the oscillator for the Propeller and is used as the master clock for all timing functions. It provides identical information to all the cogs and can be read simultaneously by all the cogs. Its major purpose is to support the timing of delays and other time-related functions. It does not keep track of the time elapsed or the number of times it has overflowed. If you need that information, you have to create the program functions to do so. All other timing functions can be implemented by observing a time differential based on the counts in the system counter at the two times of interest and knowing the operating frequency of the system—meaning that if we have two readings of the system counter and we know how fast the counter is running, we can calculate a time interval.

Program Storage and Execution

In this book, we are talking about Spin code execution. Any code examples discussed will be written in Spin.

When we move a program we wrote on the PC, in the Propeller Tool environment, to the Propeller chip, the program is moved either to main RAM in the Propeller or to the external serial EEPROM chip connected to the Propeller. Moving it to main RAM is much faster (with the Fn 10 key), and the RAM location is used for most developmental purposes because of this.

If the main RAM option is chosen (Fn 10), the program is lost when the Propeller is turned off. The program will have to be reloaded into the Propeller the next time you need to use the program. Whatever you might have stored or saved in your PC will, of course, still be available for reuse.

If, on the other hand, the program is moved to EEPROM (with the Fn 11 key), it will be available even after the Propeller system has been turned off. Downloading to the EEPROM takes much longer than downloading to the main RAM. However, programs in EEPROM are not volatile.

When the Propeller system is turned on, it looks for a connection to a PC. If there is no connection to a PC, the current program in the Propeller external EEPROM is read, moved to Propeller main RAM, and executed. Therefore, the rule is, Once the application you have in mind has been finalized, store the program to EEPROM. While you are developing the program, use RAM storage.

Spin code is always executed from main RAM, by the Spin Interpreter running in the currently selected cog. The Spin Interpreter is copied from main ROM to Cog 0 at startup, and then to any other cog as it is started by the application to run Spin code later on.

Cog_0 (an arbitrary name) can launch other cogs after startup; then after that, any cog that is running can start and stop any other cog, as needed, and even stop itself. The Spin Interpreter running in the cog fetches tokens as needed from main RAM

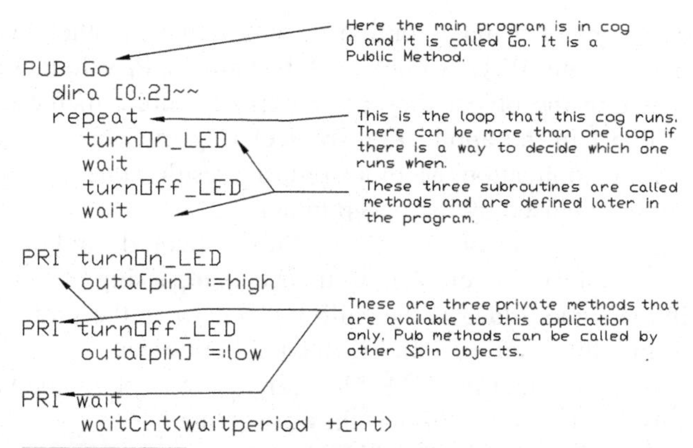

Figure 2-4 **How methods are organized and called in a Spin application**

each time that cog has access to the hub and then executes them. Each inactive cog is still assigned its time slot by the hub. This is necessary to keep everything in sync.

At least one cog needs to remain active at all times to keep the system alive. All other cogs can be turned on and off by each of the cogs and methods killed and loaded as needed to do the work to be done. Turning a cog off saves power. (Appendix C contains a demonstration program that shows you how to turn a cog on and off).

At this stage, we do not need to worry about how the hub operates, what the system clock does, or how other various housekeeping functions are handled. If you need more information on these functions, they are described in detail in the Propeller Manual. For now, we can consider them to be transparent and thus invisible to us.

A preliminary look at how a program is organized is provided in Figure 2-4.

Objects, Methods, and Other Definitions

Spin defines the components and elements that make up Spin programs as follows.

- An *application* is the collection of object files that would be the equivalent to a single-file "program" in a non-object-based language. An application usually includes a top object file and a number of other objects, but an application can be just a single object.
- An *object* is any file with a .spin extension, and it is a chunk of executable Spin code. An object may be designed to accomplish a whole application by itself, or it may be designed to interact with just a specific device and be managed by another object as part of a larger application. An application's *top object* is in control when execution begins. This top object may include an OBJ declaration listing the other objects that will be called from the top object as the application is running.

- A *method* is the equivalent of a main routine or subroutine in other languages. Methods are created as Private (PRI) or Public (PUB) entities. Private methods can only be called from within the object they are a part of. Public methods can be called by any other object in the application by declaring the object in the OBJ declarations in the calling application. Methods return a result value automatically, though the result may not be used by the programmer.
  An object can contain any number of Private and Public methods and can call the Public methods in other objects by referring to their encompassing objects. Called objects have to be in the same folder as the calling object *or* in the same folder as the Propeller Tool so that they can be found by the calling object.
- *Global variables* are defined under the VAR block and have to be defined as bytes (8 bits), words (16 bits), or longs (32 bits) as they are created.
- *Global constants* are defined under the CON block and are given names and values as they are created.
- *Local variables* are defined on the first line of the method they are used in. They are local to the method and not available outside it. They will be defined as longs.
- *Stack space* is memory (in the main RAM) that is needed by each cog to operate within. It is assigned in the VAR block as a number of longs for each cog that will be started. The startup of each cog refers to the stack space assigned to it so that each cog is allocated space without conflict with other cog space.
- *Memory quantities* are also defined in terms of bytes, words, and longs. Bits are not used as a unit of memory designation, so a memory bank would be described as consisting of 512 32-bit longs as opposed to 16 kilobits. There are no instructions for manipulating just a bit, although one bit can be manipulated in a more comprehensive instruction.

Although we are not yet ready to deal with the Spin language, I've listed Program 2-1 to get you thinking about it. This program can be run.

Program 2-1 Complete Listing of the Partial Program Shown in Figure 2-4

```
{{12 Sep 09   Harprit Sandhu
BlinkLED.spin
Propeller Tool Ver. 1.2.6
Chapter 2 Program 1

This program turns an LED ON and OFF, with a programmable delay
It demonstrates the use of methods in an absolutely minimal way.
Since the clock is 10 MHz

Define the constants we will use.

}}
CON
  _CLKMODE=XTAL1 + PLL2X           'The system clock spec
  _XINFREQ = 5_000_000             'the crystal frequency
```

(continued)

Program 2-1 Complete Listing of the Partial Program Shown in Figure 2-4 (*continued*)

```
  inv_high      =0                'define the inverted High state
  inv_low       =1                'define the inverted Low state
  waitPeriod    =5_000_000        'about 1/2 sec switch cycle
  output_pin    =27               'line the led is on

'High is defined as 0 and low is defined as a 1 because we are using an
'inverting buffer on the Propeller output

PUB Go
  dira [output_pin]~~           'sets pin to an output line with ~~
  outa [output_pin]~~           'makes the pin high
  repeat                        'repeat forever, no number after repeat
     turnOff_LED                'method call
     wait                       'method call
     turnOn_LED                 'method call
     wait                       'method call

PRI turnOn_LED                  'method to set the LED line high
    outa[output_pin] :=inv_high 'line that actually sets the LED high

PRI turnOff_LED                 'method to set the LED line low
    outa[output_pin] :=inv_low  'line that actually sets the LED low

PRI wait                        'delay method
    waitCnt(waitPeriod + cnt)   'delay is specified by the waitPeriod
```

Program 2-1 demonstrates, in the simplest of ways, the creation of three private methods that among them allow the application to turn an LED on and off and to provide a delay between the switching actions.

3

THE HARDWARE SETUP

Note *Parallax provides a Propeller Education Kit (#32305) that is suitable for almost all our experiments and for all the experiments Parallax has created for their more advanced and formal educational programs. Although it is not* strictly necessary *for us to have one, this kit is the easiest way for us to obtain almost all the components needed for the work we will undertake in one convenient package. Later on you can use this kit to undertake all the exercises provided in the Parallax courseware. Everything in the kit can be used for other purposes. (If you want to get proficient in the use of the Propeller, you will want to study and run all the educational programs provided by Parallax. The related texts are free to download!)*

We want to start working with the system as soon as possible, and in order to do that we need a hardware/software setup that we can try our hardware and software ideas on. In this chapter, we will set up the system and get it ready for our first experiment. Our setups will be based on the Propeller Education (PE) Kit provided by Parallax whenever feasible. Of course, you do not have to use this kit, but I strongly recommend that you get it. It will simplify the learning process if we are working with the same hardware. This kit does not include the Propeller Manual that you need, but you can download a copy of the manual from the Propeller Tool's Help menu at no charge. It is included as a PDF file. If you can afford it, you should get a hardcopy version of the manual from Parallax. There is nothing like having the book in your hands, and there is nothing like an electronic copy if you need to search a text file. They complement one another.

Because we need to have a standardized setup that we can both work with, we will work with the basic layout exactly as suggested for the PE Kit provided by Parallax. This will allow you to do all the experiments that Parallax provides along with almost all the beginner's experiments I have designed. All illustrations will reflect this layout, and all wiring diagrams will follow the layout suggested in the kit data. When changes are made to these layouts, they are called out in the descriptions. All the illustrations in this resource are in black and white to keep printing costs to a minimum. Identical illustrations, for most setups, in full color are provided on the support website

(Encodergeek.com) and should be referred to for greater detail. It is much easier to glean information from a full-color illustration. See Figure 3-1 for a look at what the suggested PE Kit layout looks like.

We will be using the 5 MHz crystal that comes with the PE Kit to control the frequency at which the Propeller operates. We will specify the oscillator speed for all our experiments with the following two instructions:

```
_CLKMODE=XTAL1+ PLL2X      'The system clock spec multiply factor is 2.
_XINFREQ = 5_000_000       'External oscillator frequency.(Crystal)
```

With these instructions, the Propeller will operate at 10 MHz. The frequency is very stable with a crystal (most crystals exhibit some drift with temperature and as they age). If we had used the internal RC network to specify the oscillator frequency, the frequency would not have been predictable with such accuracy. (See the Propeller Manual for a further discussion of system speed specification.)

Note *All Parallax material photographs carry Parallax copyrights and are used by permission.*

Courtesy of Parallax Inc.

Figure 3-1 **Propeller Educational Kit.**

Courtesy of Parallax Inc.

Figure 3-2 The Propeller Professional Development Board.

The PE Kit, with a few additions, will be adequate for all the experiments we will be conducting for this book. If, however, you have plans to work with the Propeller system over an extended period of time, an investment in the Propeller Professional Development Board (PPDB) offered by Parallax is worth considering. This flexible board provides a lot of accessories around its perimeter and will save you a lot of time and money over the long haul. I have provided a photograph of the board for your review in Figure 3-2.

If, on the other hand, you are a software person and all you want to do is some software experimentation and development and you will not be adding a lot of hardware, the Propeller Demo Board may be the best investment for you. It has the interfaces you need for a keyboard, a monitor, a mouse, and much more. It is ready to use (see Figure 3-3).

The Propeller chip itself is available in three form factors. They are illustrated in Figure 3-4.

Setting Up the Hardware

Set the PE Kit up exactly as suggested by Parallax. This will allow you to conduct all the more advanced experiments that Parallax offers as a part of their educational program and the more simplified beginner's experiments we will be undertaking as a part of the beginner's learning experience in this book.

The experiments created by Parallax are more formal and in many ways more suited to a first course about the Propeller at a junior college or a university.

Courtesy of Parallax Inc.

Figure 3-3 **The Propeller Demo Board.**

The experiments I have designed are more for the amateur engineer, the technician, or the hobbyist. I will not go into the scientific basis for the experiments, except at the most rudimentary level, so that we can proceed more with the use of what we are creating as opposed to the formal understanding of the science behind what we are creating. (In this I do not mean to set aside the value of the fundamental scientific understanding of all natural phenomena but rather I am committing to keep things simple as suitable for the absolute beginner. I urge you to understand what we are doing in the most fundamental scientific way possible. Many books are devoted to just that.)

Courtesy of Parallax Inc.

Figure 3-4 **The three Propeller P8X32A form factors.**

A Fundamental Reality We Have to Consider

The Propeller chip is designed to be a device that uses very little power. It runs at 3.3 VDC and is designed to operate at lower frequencies that require much less power. These features make the chip very desirable for thousands of portable applications that demand a very low power drain, but it causes some special problems for us experimenters. Among these, the one we need to address first is that a lot of the devices we will be interacting with will be operating not at 3.3 VDC but at 5 VDC and will be using TTL-level logic. The CMOS circuitry in the Propeller will switch the TTL signal without difficulty, but because we have a very limited amount of power available directly from the chip to power all the 32 I/O lines, we have to buffer the outputs from the Propeller to amplify the signals we need. (The high impedance inputs need very little power to switch them high and low, but the outputs will have to be amplified for many of the uses we have in mind.) Buffering the outputs also protects the Propeller in the case of wiring mistakes that may introduce high voltages and currents to the buffered pins. (If we do not buffer the outputs, we will be limited to loading only three or four lines that are not going to high-impedance devices.)

Let's make the reasonable assumption that for most of the experiments we undertake in this book, we will need no more than 24 lines of I/O. Of these, we will need seven lines for the 16-character-by-2-line LCD (liquid crystal display) we use as the display for all our experiments. This leaves 17 bits for other interactions. As a general rule, most applications need two inputs for every output they support, so we will need to set the balance of the 17 bits as 6 bits of output and 11 bits of input. This means we will need to provide one 7404 hex buffer for the 6 output bits. Using three 7404s would allow us to have up to 18 fully loadable outputs. The circuitry within each 7404 is shown symbolically in Figure 3-5. (A buffered line can usually drive about 10 TTL level loads.)

The 7404 hex inverter is a 14-pin device, and each one can buffer six lines. The lines are inverted as a part of the buffering process, meaning that a low is turned into a high and a high is output as a low as it goes through the buffer. For our purposes, this can be handled in the software by defining a high as 0 and a low as 1 when the constants are defined at the top of the program. Some of the high load outputs from the Propeller will be routed through a 7404 buffer. We do not need to limit the power needed by each input because the high input impedance of the Propeller inputs requires very little power. We would need a total of three 7404s buffers for all the lines we might need to use as outputs, but in this book we will never need more than one. The lines to the 16×2 LCD we will be using have very high impedances and therefore do not need to go through buffers. The use of a buffer to power and LED is shown in Figure 3-6.

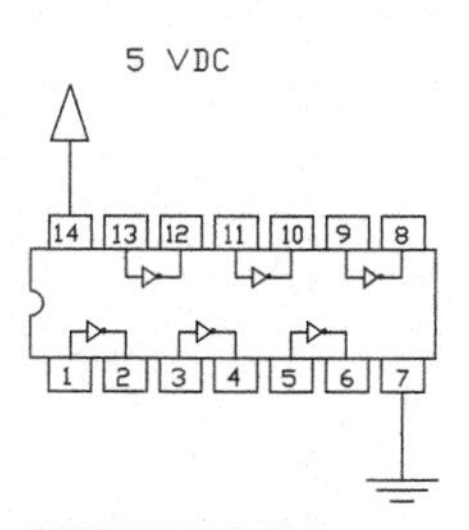

Figure 3-5 Pinouts for the six inverters in a 7404 IC

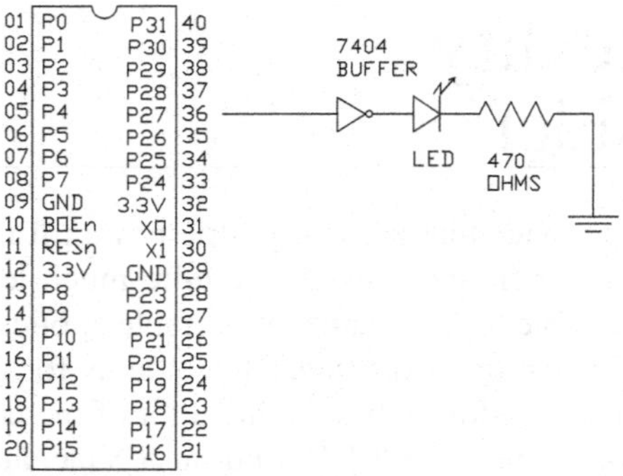

Figure 3-6 **Using one of the buffers in a 7404 to power an LED.**

In my experiments, I used the Propeller chip to turn an LED on and off (directly without a buffer) and through various resistors and then through a 7404 buffer. In doing this, I was experimenting with various conditions under which the Propeller is likely to operate. I found that even with a 1 meg resistor in line with the buffer, the LED switched without difficulty. On the other hand, no resistor is needed to manage the input line currents because the input impedances are so extremely high already.

My experiments indicated that any value of 0 to 1 meg ohms would work with the 7404 I was testing. Running mini experiments like this makes you comfortable with the operation of the devices you use every day. You will find that the inputs to the 7404 (and other similar gates) have a very high impedance so that even a direct connection will work, but that we can add a 1 meg resistor in line to make sure the load on the Propeller will be minimal, if by mistake we manage to a create a dangerous voltage at the gate input while we are setting it up. Once we get everything working the way we want it, we can remove the resistors.

The program for blinking an LED shown in Figure 3-6 is the first program covered in Chapter 4, which is devoted to the programming environment for the Propeller chip.

4

SOFTWARE SETUP: THE "PROPELLER TOOL" ENVIRONMENT

Note *In this and all other discussions, it is assumed that you have access to the Propeller Manual.*

All the programs in this book are written in the Spin language, and all programming is done in a programming environment called the Propeller Tool. This program is provided free of charge by Parallax and is available for downloading from www.Parallax.com/Propeller.

The "Propeller Tool" environment is a full-screen editor that allows Propeller programming in two languages. The first, the Spin language, is a higher-level language that will be used for writing all the programs in this book. The other language is called Propeller Assembly and consists of a comprehensive set of Assembly language instructions. All Assembly language routines must be included as DAT blocks within a Spin language program; they cannot be called as standalone methods or objects. We will not use the Assembly language.

Although the Spin language seems daunting at first sight, it is really a fairly easy-to-use language that beginners should find easy to learn after a few examples. There is really nothing intimidating about the language. The only difficult part is getting used to the *rigid indenting* required by the language for the proper formation of programming blocks. (See examples in the Propeller Manual.) This chapter introduces you to this programming environment. Extensive programming examples in the second and third parts of the book provide further examples as you get more familiar with the system.

All the programs we develop in this resource are written *exclusively* in the Spin language. The Spin language is an object-based language. It employs a formatting requirement that makes it a lot easier to use and makes it more powerful than it seems to be at first sight. If you have problems with your first programs, the most probable cause will be improper indenting. My first reaction to using the system was that it seemed more tedious than it needed to be, but then I realized it was a useful and well-thought-out way of doing what needed to be done. The Propeller Manual provides detailed instructions on how to use the language in the "Propeller Tool" environment and, of course, detailed descriptions for each of the commands. The examples given are often a little bit more sophisticated than what might be understood by an absolute beginner, and in this resource I endeavor to provide simpler examples and skip over the more advanced techniques that you find in the manual itself. My goal is to get you introduced to the system fundamentals. Once you understand the fundamentals, it will be your job to gain the proficiency you need to get your day-to-day work done.

The Propeller Manual provides no information on programming. It assumes that you are familiar with standard programming techniques, and as a matter of fact you need to understand standard computer programming techniques fairly well in order to use the Propeller Manual effectively. As we develop the programs in this book, we use simple programming techniques that are easy to understand, even for the absolute beginning user. Keeping in mind that this is a resource for the beginner, we will concentrate on the simpler, more basic techniques that need to be mastered to be able to use the parallel-processing capabilities of the Propeller system effectively. Understanding the parallel-processing environment as implemented within the Spin language is fundamental to understanding how to use this Propeller chip.

This is a new hardware environment. The basic concept that you must understand is that no matter how much hardware there is, if the software does not address the hardware features, you cannot use them. And no matter how powerful the software is, if there is no hardware to be addressed by it, the power is useless. Hardware and software must work together, and understanding how the interaction is implemented is the key to understanding how best to use any system. Because parallel processing is new to you and everything we do will be done in the Spin language, understanding the tools offered by the Spin language is critical. We need to understand how the hardware and software support one another to provide a viable environment within this system. Parallel processing needs a number of features that we are not yet familiar with. These features make the parallel-processing process possible. One example we will consider almost immediately is the need to provide a way to send up to eight separate subprograms to the Propeller's eight cogs within one program. The Spin language provides the tools needed to do this, but we don't yet understand how to use these tools. (Each cog can undertake more than one task within the subprogram assigned to it.)

In order to understand how the Propeller and its parallel-processing environment are to be used, we first need to understand a few basic concepts associated with parallel processing. The concepts themselves are not particularly alien to understanding single-thread linear programming in that each of the eight cogs is pretty much a standard linear program processor with some features added and some other features left out. The new concept that needs to be understood is that the program we write has to have some way of deciding which cog is going to do what and when so that a coherent program that can provide a viable result is created. This need is accommodated within the Spin system by providing the ability to stop and start the cogs, as needed, and passing variables and program segments back and forth between the cogs as they are created. The launching and running of the cogs is under the complete control of the programmer. He or she can start and stop any of the eight cogs as seen fit and assign whatever tasks he or she wants undertaken to each of them. The skill required to make these seemingly simple assignments is what separates the good programmers from the not-so-good programmers. I will endeavor to show you how to do this in a relatively straightforward way in this book. Once you understand the basics, you will build on what you have learned here to develop the sophistication needed to do the work required of you on a day-to-day basis.

Not only does the programmer have to assign various tasks to the cogs, he or she has to decide when to turn the cogs on and when to turn them off, which of the results being obtained need to be put into the shared memory so that other cogs can read this information when needed, and what variables are needed only locally. All the special techniques needed to make all this work have to be designed and implemented by the programmer. The system we are considering is very powerful and therefore, as can reasonably be expected, requires more skill to be used effectively.

Classroom Analogy

A classroom analogy is one way to envision a parallel-processing system. Imagine a classroom, a blackboard, eight students, and a teacher. The teacher's job is to assign each student his or her work, to call on one student at a time, and to maintain order. Each student does his or her work in another room, each with its own blackboard, in complete isolation and in silence. Students are invited to come into the classroom one at a time to read whatever is on the blackboard and to write whatever they need to on the blackboard. They read whatever they need to from the blackboard and put other variables, marks, and codes on the blackboard to provide information about the validity of the data they have put on there for the other students. The students use these marks and codes to interact with one another and to pass information back and forth via the blackboard.

The following is the case for this scenario:

- The teacher is the hub.
- The students are the cogs.
- The main blackboard is the shared memory (the hub memory, main memory).
- The students can write selected variables on the blackboard (the hub memory) and other variables on their own blackboards (cog memory, local variables).
- The students can secure areas of the blackboard with locks.
- The students can banish and reinstate other students! (Turning cogs on and off.)

There are other aspects to this analogy, of course, but basically that is how it works.

Getting Ready to Use the Propeller

We will use the Propeller Education Kit provided by Parallax, wired up as they suggest for the initial startup, connected to our IBM-PC through a USB port, working in the Propeller Tool environment.

The first thing we have to address in order to use the Propeller chip is the fact that this device runs on 3.3 VDC. The devices used in our day-to-day electronics use both 3.3 and 5 VDC. To keep things simple, we will use 5 volt TTL devices in all our experiments. This means that we need a separate power supply for the Propeller and we have to provide an appropriate interface for all connecting devices, whether they feed information to the Propeller or are being fed information by the Propeller.

The 3.3 volt requirement for the power to the Propeller chip is a blessing in that it allows us to isolate the power that the chip is using from all the other devices in our project. The easiest way to get the necessary 3.3 volts is to use a suitable regulator fed directly from the 5V power supply we are using for the rest of the project. (This is the way it is done on the Propeller Educational Kit by Parallax.)

Installing the Software

THE PROPELLER TOOL

The programming environment provided by Parallax for writing programs in the Spin and Propeller Assembly languages is called the Propeller Tool (PT). The Propeller Tool is a standalone program that provides an editing environment optimized for programming in the Spin and Propeller Assembly languages. It allows the downloading and running of the programs created (for the Propeller chip) with one keystroke. This software is designed to run on an IBM-PC or equivalent computer and is provided at no charge by Parallax. You can download it from the Internet. Almost all software is updated from time to time, so it behooves us to always use the latest version available.

The current version of the Propeller Tool is version 1.2.6. All the programs in this book were written with this version of the software. Even as I type these notes, the tool is being revised to be easier to use and more sophisticated. Download the tool and put it in the folder where you are going to store all your Spin programs. Doing this will make it easier for the software to find your programs when you need to access them for editing and running.

The tool provides a text editor and within it a very useful graphics font (it is a subset of the Parallax font) for documenting the circuitry being created. The two share the editing environment and can be used interchangeably within it. An ingenious color-coding scheme separates the various parts of the program you are writing in a logical way, making it much easier to see what you are doing. The system allows two levels of commentary and documentation to be hidden or made visible as needed. Doing it this way allows both minimal and verbose documentation to be included in the code during development and then suppressed at two levels, as necessary, to see more or less of the documentation on the screen as you develop your programs.

You can cut and paste back and forth between Word and the Propeller Tool and the formatting will be preserved. If you prefer you can write your programs in Microsoft Word, but they must be executed from the Propeller Tool screen. (It is easier to use the PT.) The font size needs to be set to 10 (or 12) to allow adequate space for remarks on each line. Set the height to match the font size. Set the tabs accordingly for each section of the Propeller Tool (under the Edit | Preferences menu).

All the programs we create will be written with the Propeller Tool, and some of the illustrations and diagrams having to do with the circuitry will be created with the Propeller font graphic facilities provided within the tool environment. Where more complicated circuitry needs to be illustrated, I provide drawings done with AutoCAD. The Propeller font does not adequately support the creation of electronic gates at this time. (Hopefully this is to be remedied in the next version of the chip. The font information is in the ROM on the Propeller chip, so it is an integral part of the chip. A new version of the chip has to be created to modify the font properties.)

The Propeller Help Menu item explains the use of the Propeller Tool in profuse detail.

No separate compiler is needed. The compiler is available automatically as a part of the Propeller Tool environment and is transparent to the user—you do not see it. When you ask the application to be downloaded to the Propeller chip, the system does so automatically and no interaction is required on the part of the user. Programs can be written in Spin, in the Propeller Assembly resource, or in a mixture of the two, and the compiler sorts everything out automatically. (All Propeller Assembly code must be listed in a DAT block within a Spin object.)

Our First Program

Install a 7404 on the education kit board and wire in one LED, as was shown in Figure 3-3 (refer to Chapter 3).

Before we get into the intricacies of the Spin language, let's write a short program that is as simple as we can make it. In this program we are going to blink an LED fed from one line of a 7404 buffer. At the same time, we will vary the resistance in the line feeding the 7404 buffer to determine the value of a suitable "in line" resistance that the circuit will tolerate and still operate reliably. We could calculate this value, but we will instead do it using the trial-and-error method. We know that the 7404 buffer has a high impedance input. This means that it looks like a high resistance device to the Propeller. Even so, we will add more resistance to this line to further limit the current drawn from the Propeller (see Figure 4-1).

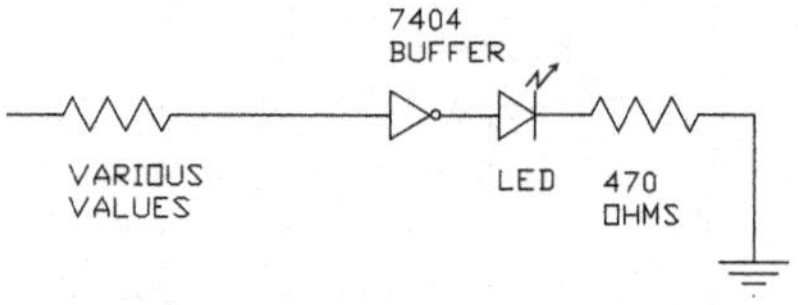

7404
BUFFER
VARIOUS
VALUES
LED
470
OHMS

Figure 4-1 **Circuitry between the output line of the Propeller and LED**

Figure 4-2 Photo of Propeller: 7404 and LED

Note *Adding this resistance will slow down the operation of the gate ever so slightly, but we may not be able to detect the change with the instruments at our disposal.*

The program we are about to consider blinks an LED approximately once a second—a half second on and a half second off—for each complete cycle with the hardware extensions shown in Figure 4-2.

We do not have to go through a buffer, but doing it this way limits the current load on the Propeller.

Program 4-1 appears in the Parallax font here. This font will be used for all programs developed and listed in this book to differentiate the program listings from the general book verbiage.

Program 4-1 Blinking an LED about Once a Second

```
Re-listed

{{12 Sep 09   Harprit Sandhu
BlinkLED.spin
Propeller Tool Ver. 1.2.6
Chapter 4 Program 1

This program turns an LED ON and OFF, with a programmable delay
It demonstrates the use of methods in an absolutely minimal way.
The clock is 10 MHz
```

(continued)

Program 4-1 Blinking an LED about Once a Second (*continued*)

```
Define the constants we will use.

}}
CON
  _CLKMODE=XTAL1 + PLL2X          'The system clock spec
  _XINFREQ = 5_000_000            'the crystal frequency

  inv_high      =0                'define the inverted High state
  inv_low       =1                'define the inverted Low state
  waitPeriod    =5_000_000        'about 1/2 sec switch cycle
  output_pin    =27               'line the led is on

'High is defined as 0 and low is defined as a 1 because we are using an
'inverting buffer on the Propeller output

PUB Go
  dira [output_pin]~~            'sets pin to an output line with ~~
  outa [output_pin]~~            'makes the pin high
  repeat                         'repeat forever, no number after repeat
     turnOff_LED                 'method call
     wait                        'method call
     turnOn_LED                  'method call
     wait                        'method call

PRI turnOn_LED                   'method to set the LED line high
    outa[output_pin] :=inv_high 'line that actually sets the LED high

PRI turnOff_LED                  'method to set the LED line low
    outa[output_pin] :=inv_low  'line that actually sets the LED low

PRI wait                         'delay method
    waitCnt(waitPeriod + cnt)    'delay is specified by the waitPeriod
```

NOTES ON THE FIRST PROGRAM

Make note of the following items in Program 4-1, starting from the top and moving down:

- If we do not specify a frequency as such for the system, the crystal frequency is multiplied by two as specified in the first two lines of the program.
- We are using a 5 MHz external crystal and a 2× multiplier for a 10 MHz operation.
- All the constants are defined up front on top of the program. This makes it easy to make changes to the constants.
- The direction of the pin being used has to be specified. This is done with the ~~ notation for output and ~ for input in the dira command.
- We are using a wait period of freq/1000 500 (or 5,000,000) cycles (half a second).

Methods are used to do the following:

- Turn the LED on.
- Turn the LED off.
- Create the wait period.

(If an LCD is attached, which should not be the case at this time, its contents will not be cleared or affected by this program.)

RUNNING THE FIRST PROGRAM

Once the program has been entered on the computer, and everything is hooked up, we can run the program by pressing the F10 key on the PC keyboard. The program will be transferred to the main RAM of the Propeller and executed. If things are right, the LED on line 27 blinks on a one-second cycle. If the crystal and its specification at the top of the program do not match, the program will not run properly. If things are close, touching the crystal might add enough capacitance to the system to make things work. If the blinking does not start spontaneously, and reliably, things are not right. Before we go any further, we must get this right. Look over the hardware and the software to see what is not right—and then fix it.

Keep this program handy for checking the operation of the system whenever you create a new setup and want to make sure it has the potential to work.

The Typical Spin Program

A typical Spin program is divided into six main sections or blocks that define the following components in the program:

- **CON** Constants are those values that never change in the program.
- **VAR** Variables change their values in the program over time.
- **OBJ** Calls to other objects (programs) that have methods we are interested in incorporating into our application.
- **PUB** Public routines or methods that can be called from external objects.
- **PRI** Private routines or methods that can only be called from within the parent object.
- **DAT** Data used in the program. (This is also used in Propeller Assembly programming, but we will not cover that aspect of DAT use in this book.)

Each section can have any number of lines of code in it, and each line can be commented as extensively as you like. Two types of multiline comments are supported. Those enclosed in single brackets and those enclosed in double brackets as shown here:

```
{{Comment type 1}}
{Comment type 2}
```

You can also comment each line with a single quote marker.

The Propeller Tool environment lets you list the program under consideration in four formats. The four choices appear on the top menu line of the program listing:

- *Full Source mode* lists everything that was typed in as a part of the program. All the code is listed along with all the comments.
- *Condensed mode* lists everything except the comments that are in double brackets: {{ ... }}. For us, this means that we should put all the verbose documentation in double brackets. This would include all the general comments as well as detailed descriptions of how the code works. Credits and general licensing comments should be included within double brackets.
- *Summary mode* lists the method headings under each of the main sections.
- *Documentation mode* lists general information in the double brackets and variable space used by the program. Information like how many variables are used by the program and how long the program is.

  As your familiarity with the program structures increases, you will settle on your own rules for documenting your programs. In the interim, the following rules will suffice.
  - At the top of the program, use the double brackets to extensively document what the program does in some detail along with author information and relevant dates and revisions.
  - Document each line of code with the (') marker comments so that each line is easier to understand. Explain what each line of code is doing. Where necessary, provide verbose documentation for difficult-to-understand lines.
  - Look at all four formats occasionally as your program evolves to see what is being listed under each format and adjust your commenting to make the best use of the formatting features. All the programs I provide will follow these guidelines with double bracketed comments being added to explain the use of particularly difficult-to-understand code segments.

I like to do my programming in Full Source mode most of the time and Condensed mode some of the time. The effect of using the preceding rules is reflected in how Program 4-1 is laid out. Subsequent programs will follow these rules.

The Propeller Tool allows you to have any number of programs opened at the same time. You can cut and paste freely between the programs as you develop the code for your current program. This can save you a lot of time and allows you to reuse lines of code that you know work in the way they are intended to. The names of all the open programs are listed on the top menu line of the Propeller Tool screen.

A detailed description of the Propeller Tool (PT) environment and its effective use is provided under the Help section. The PT software provides a whole host of powerful tools you need to study and become comfortable with. An extensive section consisting of a detailed tutorial on using PT is also provided in the Help section. It covers the special features of the software that aid in more rapid development of software.

The Propeller Manual contains detailed descriptions of all the commands in the Spin language and of the Assembly language commands.

We will not go over the command descriptions in this book but will cover the use of the more commonly used commands so that you will be comfortable with their use.

A more extensive use of the commands is developed in the programs in the second and third parts of this book. The programs are devoted exclusively to the development of techniques that do real work in the real world.

Once you get comfortable with the general layout of a Spin program, you can start adding more sophisticated and complicated commands to your programs. The basic goal is to get comfortable with the Spin language and how it works. Adding little complications as you get better at programming in Spin is relatively easy once you start writing programs that work.

Program Structure

What does a Spin program look like, and where do we put what to make it work right? First, let's look at a couple diagrams to see what a linear program looks like as compared to a program set up for a parallel environment. Figure 4-3 shows a schematic for a typical linear program.

Notice that there is only one main loop. Figure 4-4, on the other hand, shows what the layout for a parallel programming schematic might look like. Notice that a number of independent programs share a common memory bank. The only interaction between the programs is through the shared memory. This is the hallmark of the parallel programming environment provided by the Propeller system.

Note *Each processor has its own loop, but they all use the shared memory.*

The difference is readily apparent, so next let's look at what a simple Spin program looks like as actual program code. (No parallel processing yet.)

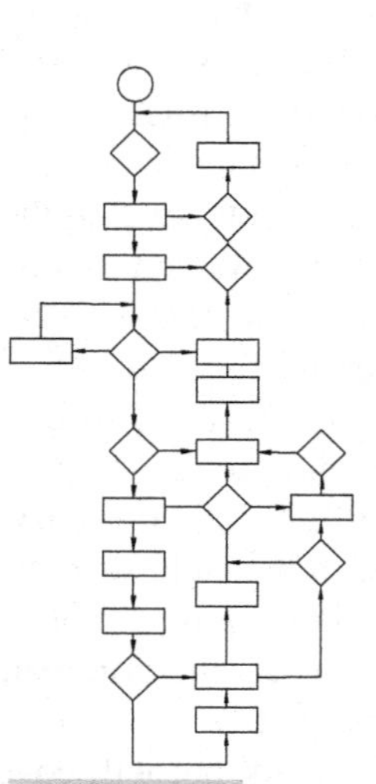

Figure 4-3 Linear program schematic

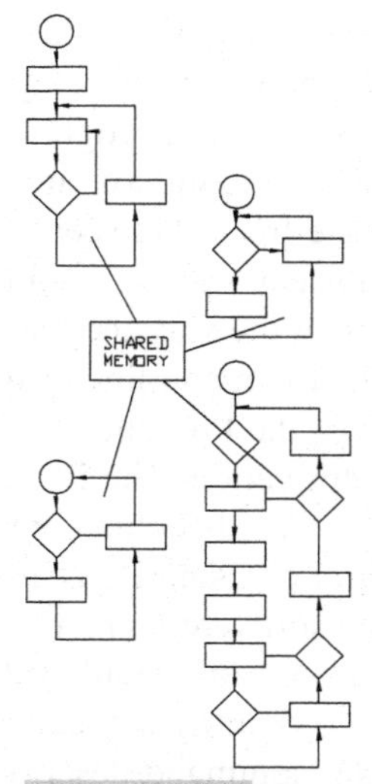

Figure 4-4 Parallel program schematic

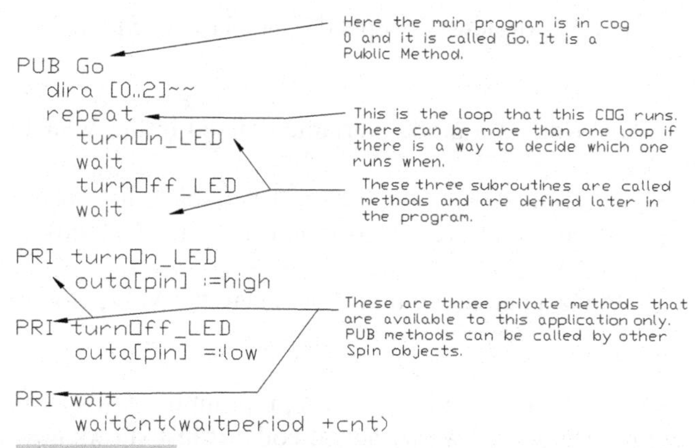

Figure 4-5 Code descriptions for the program to blink an LED

Figure 4-5 shows the code for blinking an LED, as was done in Program 4-1 earlier, with comments to explain what goes where within the object. You should also be looking at the entire code for this object (shown earlier) to see what the commenting for each line looks like and states. All objects that use only one cog are expansions of this object. The methods called are part of the object in the same cog. No methods are called from an external object in this example.

When stack space has to be assigned for another cog that will execute Spin code, the spaces are declared on top of the program in the VAR section, as shown in Figure 4-6.

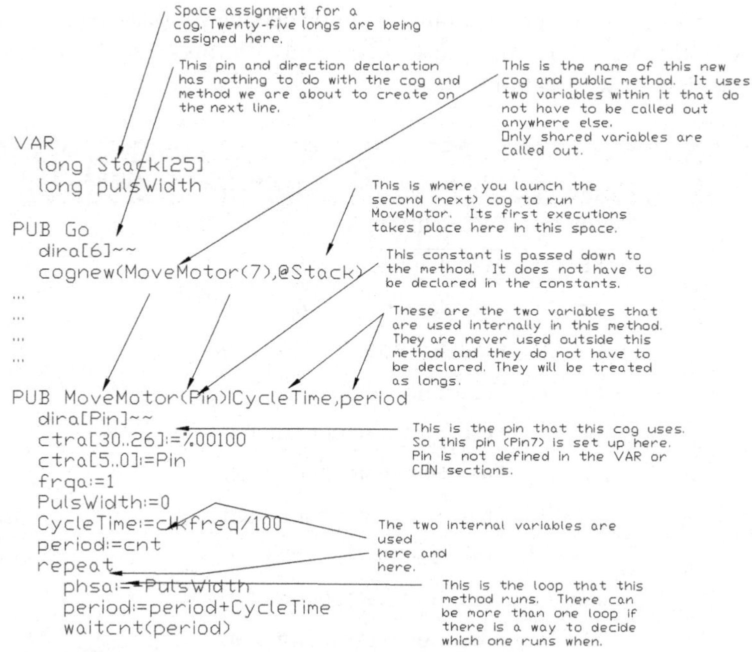

Figure 4-6 Assigning space for a new cog

This figure contains a number of new concepts that need to be explained before you look at the code:

- **Global variables** PulsWidth is a global variable defined in the VAR block. It is available to all the cogs.
- **Local variables** Cycle_time and period are local variables that are used within the MoveMotor method only and are not available to other methods or to other cogs.
- **Parameter passing** Pin is a parameter that is passed to the MoveMotor cog when it is started.

Figure 4-6 shows where all the variables go for a program that uses two cogs. The main cog in the program does not show any actual code. It just shows how the main program assigns space and then starts a cog to run the MoveMotor method. The space is assigned under VAR as 25 longs at the location Stack.

General Pin Assignments Used in the Book

In general, the 40 pins on the Propeller will be given the assignments outlined in Table 4-1. If this scheme is followed, some items can be left attached at all times and others that must be relocated have to be moved only occasionally. The information in this table forewarns you about what pins might be needed for our experiments so that you can assign the pins you want to use accordingly. In this table, I = input and O = output for the data flow directions.

TABLE 4-1 PIN ALLOCATIONS FOR THE PROPELLER CHIP AS USED IN THIS BOOK

PIN	USAGE	DIR	DESCRIPTION	COMMENT
P0	As needed.	I or O	Encoder when used.	
P1	As needed.	I or O	Encoder when used.	
P2	As needed.	I or O	Free. Encoder 2.	
P3	As needed.	I or O	Free. Encoder 2.	
P4	As needed.	I or O	Free. Servo/stepper motor.	)
P5	As needed.	I or O	Free. Servo/stepper motor.	)
P6	As needed.	I or O	Free. Servo/stepper motor.	) Xavien
P7	As needed.	I or O	Free. Servo/stepper motor.	) Amplifier
P8	As needed.	I or O	Free. Servo/stepper motor.	)
P9	As needed.	I or O	Free. Servo/stepper motor.	)

TABLE 4-1 PIN ALLOCATIONS FOR THE PROPELLER CHIP AS USED IN THIS BOOK (*CONTINUED*)

PIN	USAGE	DIR	DESCRIPTION	COMMENT
P10	Free.	I or O	Free.	
P11	Free.	I or O	Free.	
P12	LCD connection.	O	Dedicated permanently.	
P13	LCD connection.	O	Dedicated permanently.	
P14	LCD connection.	O	Dedicated permanently.	
P15	LCD connection.	O	Dedicated permanently.	
P16	LCD connection.	O	Dedicated permanently.	
P17	LCD connection.	O	Dedicated permanently.	
P18	LCD connection.	O	Dedicated permanently.	
			LED indicator 10.	
P19	Potentiometer Sel.	I or O	Dedicated semi-permanently.	
			LED indicator 9.	
P20	Potentiometer Clk.	I or O	Dedicated semi-permanently.	
			LED indicator 8.	
P21	Potentiometer Dout.	I or O	Dedicated semi-permanently.	
			LED indicator 7.	
P22	Potentiometer Din.	I or O	Dedicated semi-permanently.	
			LED indicator 6.	
P23	Free.	I or O	Free. Goes through buffer, LED indicator 5.	
P24	Free.	I or O	Free. Goes through buffer, LED indicator 4.	
P25	Free.	I or O	Free. Goes through buffer, LED indicator 3	
P26	Free.	I or O	Free. Goes through buffer, LED indicator 2.	
P27	Free.	I or O	Free. Goes through buffer, LED indicator 1.	
P28	System usage not to be disturbed by us, but can be used as I/O.			
P29	System usage not to be disturbed by us, but can be used as I/O.			
P30	System usage not to be disturbed by us, but can be used as I/O.			
P31	System usage not to be disturbed by us, but can be used as I/O.			

The pin allocations were decided upon based on the following considerations. If we agree on these assignments for our experiments, the disruptions between experiments will be minimal and the methods that we develop will be able to be called from third-party programs without worrying about what is connected to what pin for each experimental setup.

The need to read an encoder is accommodated by assigning the first two pins as "quadrature encoder information" capture pins. We will use only one encoder later in the book to control a DC motor. It is often the case that two motors are needed to accomplish even simple tasks. We will provide for two encoder inputs up front. Pins P2 and P3 can be assigned to this. If encoders are not connected, the pins can be used for input or output. Input is more appropriate for these pins because no buffering is provided for output loading.

The two-axis amplifier we will be using to run the servo and stepper motors later on in Part III of the book needs to be connected to the Propeller with six pins. Pins P4–P9 have been assigned to this usage. If an amplifier is not in use, these pins can also be used for any input or output application. No buffers are provided on any of these lines.

The LCD will be connected to the Propeller initially with an 8-bit path for the data and later with a 4-bit path for the data. It needs three lines for data transfer, chip select, and so on, so a total of 11 lines have been assigned to this use. Only lines P12–P18 are needed if four-line LCD control is implemented. When not being used by the LCD, these lines can be employed for any other use. As with the other lines mentioned earlier, no buffering is provided (or needed) on these lines.

Lines P19–P22 have been assigned to reading the two potentiometers used often in the experiments to read in data. They can use either the MCP3202 A2D converter to read two lines or the MCP3208 to read eight lines. These lines can be used for other applications when needed. (These chips read a voltage from 0 to 5 volts to a resolution of 12 bits very rapidly.)

A 7404 hex buffer will be used to provide buffering for signals that need more power than what's provided by the bare Propeller lines. Lines P22–P27 can be used as needed, with or without the six buffers provided by the 7404.

A more detailed description of the possibilities with a complete wiring diagram is included in Appendix D, along with a comprehensive board design suitable for extensive experimentation. It follows the preceding guidelines and implements the use of connectors and jumpers to make the use of the board as convenient as possible. All the experiments in this book can also be performed on this board, and later on you can use the board for the many experiments you design.

Propeller FAQ*

What is special about Cog 0?

Not much. It is exactly like all other cogs except that it happens to be the first cog started when the Propeller chip boots up. See the boot-up procedure. So, all applications begin with Cog 0, executing the first method in the top object. After that, the

*Reprinted with permission by Parallax Inc.

programmer can determine which cog will execute which portions of the application. All cogs are physically identical.

Can Cog 0 be turned off?
Yes. Upon startup, Cog 0 is loaded with the Spin Interpreter and starts fetching and executing application code tokens from main RAM. This application code can include a command to start a Spin or Assembly process in another cog with COGNEW or COGINIT and then shut itself down with COGSTOP(0). Cog 0 can be restarted by code running in another cog.

Can any cog start or stop any other cog?
Yes, by using COGSTOP and specifying the target cog ID. A common strategy when starting another cog is to store its ID for use in a future COGSTOP command when it's time to stop it.

Can a cog shut itself down?
Yes. In Spin, any cog can simply use the COGSTOP(COGID) command in a method. In Assembly, it may look something like this (assuming MyID is an available register):

```
Cogid MyID        'get our cog id
Cogstop MyID      'terminate this cog
```

Do I have to indent the lines under PUB or PRI for those lines to be part of the method?
No. Indenting the lines that are part of the method just makes it easier to read. Indenting *is* necessary to define other types of code blocks, such as REPEAT and IF.

What do I have to write in my code to indicate the end of a method?
Nothing! You do not have to end a method with any specific command or with a blank line. The compiler knows that a method is complete when it finds the beginning of the next block declaration (CON, DAT, OBJ, PRI, PUB, or VAR) or when it reaches the end of the program listing. In sample code listings, you will probably see one or two blank lines between methods just because they are easier to read that way.

Can a cog launch an assembly routine into itself?
Yes, it can, using COGID to identify itself (if the cog is not known) and COGINIT. If the cog was initially running Spin code, the Spin Interpreter in cog RAM will be overwritten with 496 longs from main memory, starting with the address where the desired assembly routine begins.

What do I have to write in my code to indicate the end of an Assembly routine?
Most Assembly routines are infinite loops and don't need any end-of-routine indicator. In the rare case where you need to make an Assembly routine terminate at the end of its operation, you need to instruct it to shut down the cog that is running it. You can do this with the Assembly versions of COGID and COGSTOP. It may look something like this (assuming MyID is an available register):

```
Cogid MyID        'get our cog id
Cogstop MyID      'terminate this cog
```

If you do not do this, you may run into strange behavior, and here's why: When an Assembly routine is launched, the cog fills its RAM with 496 longs from the main RAM, beginning at the AsmAddress specified in the COGNEW or COGINIT command. If your routine is less than 496 longs, the cog will also be "slurping up" whatever is in adjacent memory, which could be data, variables, or even another Assembly routine from the same DAT block. If your Assembly routine is not an endless loop, and it does not terminate by identifying its cog and shutting itself down, at the end of the routine the cog will keep going and try to execute this "slurped-up" data. This usually results in undesirable, and sometimes unpredictable, behavior.

Since the Propeller has eight cogs, does this mean I can have up to eight object files in my Propeller application?
No! There is no direct relationship between objects and cogs. Applications are limited by the size of the Propeller chip's main RAM, which is 32KB (kilobytes), and not by the number of object files that make up the application. An application may consist of a single object, or many objects, as long as the total size of the application is less than 32KB. An application, whether made from one object or many objects, may execute with one, two, or up to eight cogs, depending only on the collective objects' "requests" to launch cogs.

So how many processes can the Propeller chip handle at once?
You can have up to eight processes executing at any one time. This does not limit your application to eight objects, eight Spin methods, or eight assembly routines, just eight processes executing code at the same time. Some processes may need to perform continuously, such as the main program loop or code that is parsing a constant stream of data. Other processes, such as those checking a slow-changing sensor or updating a message displayed on a monitor, may only need to happen once in a while, and when these are done, the cog's resources can be freed up for another process. Each cog also has two counter modules that can each handle a separate high-speed repetitive process in 32 different operation modes, monitoring or controlling up to two I/O pins each.

Where does my application code live on the Propeller chip?
The application you write in the Propeller Tool will reside, in its binary form, in the Propeller's main RAM, not in cog RAM. Spin code is executed by the Spin Interpreter, running in a cog's RAM, which fetches and executes chunks of code, called tokens, from the PUB and PRI blocks of the application in main RAM. Assembly code is actually loaded from the application in main RAM into cog RAM and executed directly, which is why it is so much faster.

How do the cogs run Spin code? What does the Spin Interpreter do?
Once the Propeller chip has run its boot-up procedure, the Spin Interpreter stored in the main ROM is copied to Cog 0's RAM. This Spin Interpreter fetches chunks of the application code, called tokens, from the main RAM. Execution begins with the first method in the application's top object. Cog 0 fetches one or more tokens, executes the related code, then gets more tokens and continues. Whenever the application launches a new cog with COGNEW SpinMethod or COGINIT SpinMethod, that new cog also

gets a copy of the Spin Interpreter in its own cog RAM. The new cog then starts fetching and executing tokens from the application in main RAM too, starting at the point indicated in the SpinMethod argument of the COGNEW or COGINIT command. So, at any given time, there can be up to eight cogs using their own copies of the Spin Interpreter to fetch and execute tokens from the application.

How do the cogs run Propeller Assembly code?
Once the Propeller chip has run its boot-up procedure, the Spin Interpreter stored in the main ROM is copied to Cog 0's RAM. This Spin Interpreter fetches chunks of the application code from the main RAM and executes it. When that code is a COGNEW AsmAddress or COGINIT AsmAddress command, the Propeller starts the designated cog, and that cog loads its cog RAM with 496 consecutive longs of data from main RAM, starting at the AsmAddress location. (This overwrites any code or data that may have been in the target cog's RAM.) The cog then executes the Assembly code directly.

Can I launch a new cog from an Assembly routine?
Yes, you can, using the Assembly version of COGINIT. You can specify a target cog by ID, or request the next available cog.

How long does it take to launch an Assembly routine into a cog?
Approximately 512 × 16 × (1/clock speed) seconds.

How are Spin methods similar to subroutines in other languages? How are they different?
Spin methods are like subroutines in that they contain the instructions to perform a specific process, and can be used over and over again as needed. Unlike subroutines in languages such as BASIC, methods can be passed parameters—that is, a set of temporary values used in the execution of that instance of the subroutine—without having to define variables or constants globally in the program.

What's the difference between global and local variables?
Global variables are declared with a VAR declaration and are available for use by all the methods within that object. Local variables are declared in the method declaration and can only be used in the method where they are defined. All variables, whether global or local, exist in main RAM; "local" does not mean the variable exists only in a certain cog's RAM.

5

THE VARIOUS PROPELLER MEMORIES

Note *The programs that the Propeller runs may be stored permanently in an I^2C external EEPROM, but not inside the Propeller. If you do not include an EEPROM chip in your design, you will not have a standalone system. The Propeller will have to be loaded with your program every time you want to use it if there is no EEPROM or if there is no program in the EEPROM.*

As you can well imagine, it takes more than one memory bank to run a parallel processing system. How does the Propeller system organize its memory banks, and how are they used? What do we have access to and what is the best way to set up our programming strategies? This and other related ideas are introduced in this chapter.

We can better understand what we are talking about in this chapter by taking a look at Figure 5-1. Study this figure for a few seconds to get a firm idea of how the overall system is set up before we start talking about it.

We need to be aware of a number of memory banks and special memory locations in the Propeller chip so that we can program it in the most efficient way possible and with as few memory-related problems as possible. Most of the memory functions are handled automatically by the nature of the architecture, and we as beginners don't

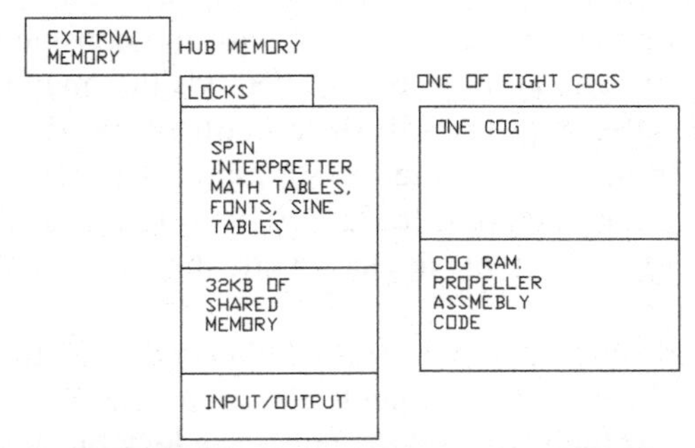

Figure 5-1 An overview of the Propeller memory banks

have to worry about what is happening where, or why. However, knowing a little bit about what is going on in the system will help you to be a better programmer.

As you read this chapter, keep in mind that *all eight cogs are absolutely identical* in every way. Upon startup, Cog 0 is started and begins executing the first method in the top object, but after that all cogs are equal.

The important thing you need to keep in mind, as you read, is the assigning of memory space for new cogs as they are launched to execute Spin code. This is not overly difficult at your current stage of the learning process—for now we can just over-assign the amount of space needed and things will be okay. Once we get the program running, we can lower the amount of memory assigned to each cog a little bit at a time until the program starts to malfunction. Then we can back off a couple longs and everything will be fine. Once the code is set, you can use an object in the Propeller Tool (PT) for determining stack space needs more accurately. Things get more complicated when you start writing rather long programs (and start turning cogs on and off and keeping track of what is going on in which cog). As far as opening a new cog goes, asking for a new cog is all you need to do because they are all identical, so it does not matter which one starts next.

Two KB of dedicated RAM is organized as 512 32-bit longs in each cog, called cog RAM. This is the cog's own "personal" RAM. This is where the Spin Interpreter is copied to from main RAM (or where Propeller Assembly code is copied to and executed) and for immediate use within the cog for internal calculations and dedicated special-purpose registers and general programming requirements. Only the cog this RAM is in can access this RAM, meaning that the cogs are insulated from each other. The only exception is the ability of one cog to pause or shut down another cog and to specify the program that the new cog is to run.

All the cogs share an external 32KB of hub RAM, which is addressable as bytes, words, or longs. For our purposes, the basic use of the shared RAM is to allow the cogs to communicate among one another by sharing access to variables each of them create in this space and to hold the downloaded application. All shared variables are declared in the VAR section of this memory. If you want to share a constant, declare it in VAR and then set it to the value you want to share (from time to time, if it changes).

Thirty-two input/output (I/O) pins can be accessed by all eight cogs at all times. The first cog started controls the initial clock frequency. The clock speed–related registers are set in this cog at the top of the main program, along with the constants. These values are used by the main program and all objects and methods that may be called from the main program. Even if a called object sets its own oscillator speed, that speed will be overwritten by the initial frequency specified at startup in the first cog. After startup, any active cog can change the operating frequency. Setting a low clock frequency is an easy way to save energy in portable devices while they are in an inactive or dormant mode.

Eight bits are used as locks to prevent memory access conflicts between the various cogs. These global bits are accessed with hub commands and are available to each cog when it has control of the system (during its turn on the hub). The lock bits are defined by the system and are used for multiple cog access control of shared memory. If more than one cog is writing to more than one 4-byte long in the shared memory at one time, the possibility exists that an access conflict will occur if the hub transfers control to the next cog in the middle of a transfer—meaning that there is a possibility of reading and writing to the shared memory in a conflicting way. The lock bits allow us to control the read/write interactions and ensure orderly program flow. (See the related discussion in the PT Help section for more information on this. We will not use the lock bits in any of the programs in this book.)

The main memory (not the cog memory) consists of a total of 64KB of memory divided between 32KB of RAM and 32KB of ROM.

All the cogs share the block of 32KB of main ROM. This ROM contains the Spin Interpreter, the mathematics support tables, the bitmaps for the characters for the Parallax font, and so on. For all practical purposes, we as beginners do not have to worry about this part of the memory at this time—its operation is transparent to us. The interpreter is downloaded into each cog. When it is the cog's turn, its interpreter fetches tokens from the main RAM and executes them.

The 32KB section of shared RAM on the Propeller chip is accessible (shared) by all eight cogs. If a cog needs to bring something to the attention of another cog, the relevant information has to be placed in this area of the memory. If certain flags are needed to alert a cog about a change in memory, it has to be placed in this part of the memory. Any information that has to be shared or accessed by more than one cog has to be placed in the shared RAM under VAR.

Assigning Memory for a New Cog

When you want to launch a new cog, you have to make an estimate of how much stack space (in main memory) it will need and then assign an estimated number of longs that represent the operational space for the new cog. At our current level of programming, we can say that five to ten longs are enough for short methods and that 30 to 35 will handle a good-sized method. None of the methods in this book take up more than 50 longs. All the stack space we assign for the various cogs is in main RAM. The memory for a new cog is assigned as shown in Figure 5-2.

A New Cog Can Be Started to Run a Private or Public Method

Once you understand the basic information in Figure 5-2, take a look at Figure 5-3, which shows an expanded version of how a new cog is launched and how the various variables play out.

For more examples of the opening of cogs, see the program listing in Parts II and III of this book. The programs tend to get progressively more difficult, so it is important to start with the shorter, simpler programs if you are having difficulty understanding the code. It is important to understand exactly what each line of code does before you move on to the more complicated code.

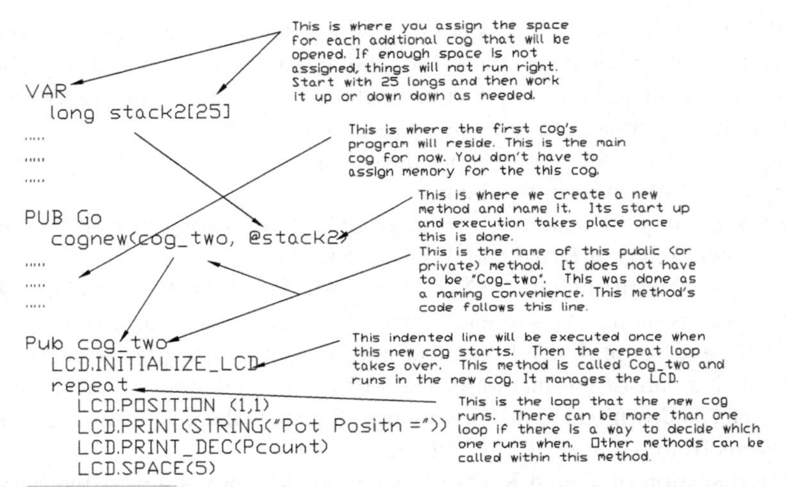

Figure 5-2 Simple cog launch explanation, with no variables

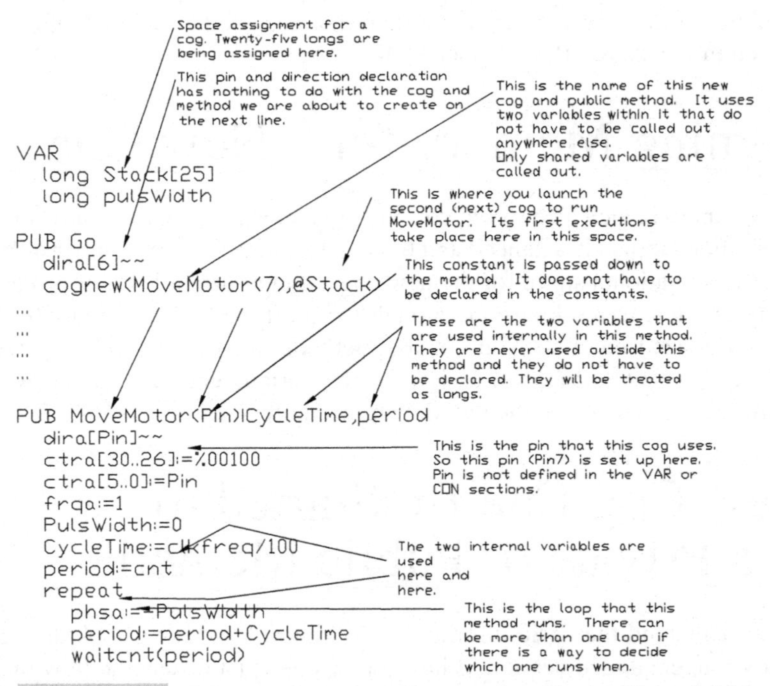

Figure 5-3 Advanced cog creation explanation, with variables.

6

THE HOW AND WHY OF SHARED MEMORY

The parallel processing environment has some special requirements as regards the memory provided for the system. We can reasonably expect that there will be a need for some shared memory that each of the processors in the parallel environment can access as needed. We can also reasonably expect that there is some memory assigned to each cog that does not need to have any input from any sources outside the cog. That is exactly how the memory is organized within the Propeller system. The main hub memory is shown diagrammatically in Figure 6-1. This is the application memory—the memory that will contain the variables, stacks, data, and code for the program you write. The first half of the memory is RAM, and the last half is ROM.

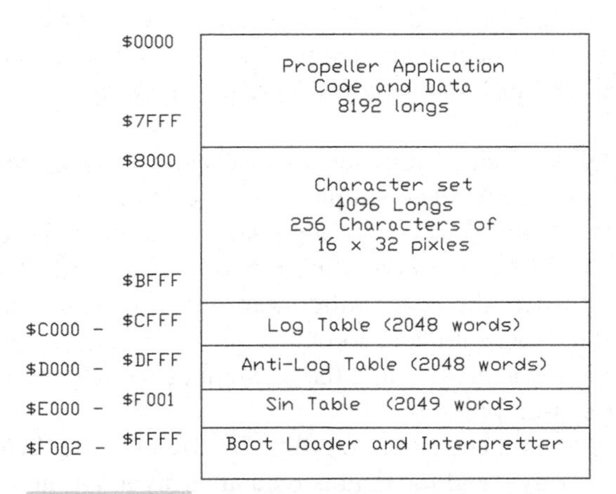

Figure 6-1 Propeller hub main memory map (from page 31 of Propeller Manual [Ver. 1.1])

Memory Usage

The program we write resides in main RAM in the hub. On startup, each cog downloads the interpreter from the main ROM to its RAM. When it is the cog's turn to access the system, it fetches tokens from its part of main RAM, as was assigned in the stack space for it, and executes these tokens in its memory space. In this way, all the cogs can be executing their part of the code at the same time. All the cogs have access to all the I/O lines at all times.

All the variables we declare in the VAR block of the program listing are available to each of the cogs whenever they have access to the shared memory. The variables can be divided into two families as static and dynamic variables.

STATIC VARIABLES

If the selected variable can be considered to be valid at all times or if our interest is in the variable "at the time when we read it" (whenever that may be), the variable can be considered *static*. This would be true for a variable such as the temperature we were maintaining within a reactor vessel in a chemical plant. It does not particularly matter when exactly we read the variable because we are interested in the current temperature only. No handshake or other confirmations are required for reading variables that meet this requirement.

DYNAMIC VARIABLES

A *dynamic* variable is a variable that changes from time to time and our interest is in the time history of this variable. We may want to know what the elevation of a rocket was for every millisecond of its assent. For such variables, it is necessary to set some sort of protocol that will make sure each data point was read at the appropriate time. Which cog does what at what time is decided by the programmer, so the storage of the data points is assigned as a critical task to one cog. The data is read by the other cogs as time allows.

For data that changes slowly (say, once a minute), we can set up a simple handshake arrangement to transfer the information. The simplest scheme involves using a flag. Cog_one places the data and sets the "data valid" flag. Cog_two reads the data and then clears the "data valid" flag. The first cog waits for the "data valid" flag to be cleared, replaces the data, and resets the "data valid" flag. As soon as the flag is set again, the second cog reads the data again, and so on.

Other variables have mixed characteristics and the techniques for reading them are mixtures of what we have discussed previously.

If a variable of interest is changing extremely rapidly, the cog creating the data has to store the data in one or two arrays, and the data is read an array at a time (alternating arrays if necessary) as system resources allow time to get the job done. Appropriate flags are created and used as was done with single data points.

Variable Validity

The variables that we use have to be created and modified in such a way that they are always current. Let's look at an example. Suppose we want to read the value of a variable (X) for use in two cogs, where one cog determines X and the other cog displays it. In order for X to be valid at all times, it cannot be used in any calculations of formulas other than the one that sets it, and it has to be set in such a way that it is always valid. Suppose the value of Y was determined as the square of a number plus 1. You cannot write this code as the following:

```
X=2
Y=X^^2
Y=Y+1
```

Because Y can be either X^^2 or X^^2 + 1 at any one time, we have no way of knowing when a cog is going to read the value because we are not using a "data valid" flag handshake. We have to code this as follows:

```
X=2
X=X^^2
Y=X+1
```

Then, no matter when another cog reads Y, it will be valid. Although this is a rather trivial programming example, variations of this error will cause you many problems if you are not following the validity rule. As a general practical rule, a shared variable may be set once and only once (in the relevant loop) by the cog that controls it. No other cog may manipulate the value of the variable, meaning that it is to be considered a "read-only" value by all other cogs.

This example also explains why more than one cog cannot be allowed to manipulate a variable. Answers using a variable that can be used in intermediate calculations will inevitably be incorrect because there is no way of guaranteeing when the variable will be valid (unless you are using locks).

Another common conflict has to do with trying to use lines that are being used to control the LCD for other purposes. Although this can be done, it is best that you not. Keep in mind that the LCD is often called from within other programs. In the other programs we forget that the LCD is actually connected to the Propeller, and if we are constantly updating the LCD it is connected with some very busy lines. These lines cannot be used for any other purposes. This rule also applies to all pins that are dedicated for use by common utilities.

Make a list of all the Propeller lines used by your program and by all the methods you are going to call from your program. The lines used in the called methods may not be used for other purposes. Table 4-1 (in Chapter 4) shows how the lines are used in the programs we are developing in this book. You may want to post a copy of that table where you can refer to it as you develop your programs.

Loops

Almost every cog sets the parameters (such as I/O line specifications and initial variable values) it needs for its operation and then launches a loop. Loops are used to perform the following tasks, for example, among a myriad of other uses:

- Update the LCD.
- Read the potentiometers.
- Make calculations.
- Read sensors.

Loops can be called once, any number of times, or forever, depending on the method's needs. Loops can be nested, and they can perform more than one task.

It is the responsibility of the programmer to make sure that the loops do not crash. Special attention has to be given to this because the Propeller does not provide safeguards against division by zero, counter underflows and overflows, register underflows and overflows, and other mathematical transgressions. This means that it may be necessary to provide checks on the ranges of variables and provide clamping within minimum and maximum values. Integer math can be negatively affected by the answer to a calculation when the order of the mathematical operations is picked without thought to register under- and overflows. Although 32-bit numbers are huge, this can still cause problems. Iterations that seem harmless when executed a few times can wreck havoc when executed infinitely.

All cogs must also create and manipulate variables in such a way that they are guaranteed to be valid at all times and therefore can be used by other cogs without having to confirm their validity. If you start to get jittery LCD displays or motor operations in the programs we develop later in the book, chances are good some invalid data resides in a loop in one of the cogs.

When a cog is started, all its variables are set to zero. This can cause problems in circumstances that depend on the values of variables to be nonzero. For example, the waitcnt instruction is sensitive to this and can be compromised. This situation can be avoided by setting variables to the appropriate values within the cog at the top of the code for the cog.

7

UNDERSTANDING ONE COG

In order to formalize the discussions about the object-oriented Spin language, Parallax has defined some concepts that we need to be familiar with before we proceed. The definitions of these terms, as used by Parallax, first appeared in Chapter 2 and are repeated here:

- An *application* is a collection of object files that would be the equivalent of a single-file "program" in a non-object-based language. An application usually includes a top object file and a number of other objects, but an application can be a single object.
- An *object* is any file with a .spin extension, and it is a chunk of executable Spin code. An object may be designed to accomplish a whole application by itself, or it may be designed to interact with just a specific device. It can also be managed by another object as part of a larger application. An application's *top object* is where execution begins. This top object may include an OBJ declaration listing the other objects that will be called from the top object as the application is running.
- A *method* is the equivalent of a main routine or subroutine in other languages. Methods are created as private (PRI) or public (PUB) entities. Private methods can only be called from within the object they are a part of. Public methods can be called from any other object in the application by declaring the object in the OBJ declarations in the calling application. Methods return a result value automatically, although the result may not be used by the programmer.
- An *object* can contain any number of private and public methods and can call the public methods in other objects by referring to their encompassing objects. Called objects have to be in the same folder as the calling object *or* in the same folder as the Propeller Tool so that they can be found by the calling object.
- *Global variables* are defined under the VAR block and have to be defined as bytes (8 bits), words (16 bits), or longs (32 bits) as they are created.

- *Global constants* are defined under the CON block and are given names and values as they are created.
- *Local variables* are defined on the first line of the method they are used in. They are local to the method and are not available outside it. They will be defined as longs.

Before we start talking about cogs, we need to keep a thing or two about a Spin program in mind. A Spin program contains the source code for each of the to-be active cogs within it. On startup, the initial cog assigns the code to the various other cogs, as specified. Once the program starts, the individual cogs are autonomous and can turn each other on and off. There are no interrupts in the Spin system, so cogs are assigned various, what would normally be, interrupt-driven tasks. A cog can undertake either one or more than one task, depending on how rapidly the tasks need to be completed. Because there are no interrupts in Spin, all the programming in a cog is linear.

Parallel processing is the holy grail of the computer industry, and the Propeller chip provides an exciting opportunity for us amateurs to investigate the possibilities. Although it is, of course, possible to do everything with a single processor, it is a lot easier if you have a number of processors available to do the work in a parallel environment. Each of the eight processors in the Propeller is capable of acting like the usual single-processor microcomputer and can undertake all the tasks you would expect a microcontroller to be able to execute. (Again, interrupts are not provided in the Propeller system, meaning that we cannot stop in the middle of a task and take care of something critical that comes up. Everything has to wait its turn.) The more pressing problem we face in small systems such as the Propeller is the constraint imposed by the limited amount of memory provided with each cog. It inhibits the handling of large amounts of data and the inclusion of standardized math/trig/log packages.

A Propeller has an external EEPROM attached to it that provides memory for program storage. It does not provide random access to the external memory's contents, however. This limitation is imposed by the fact that most serial one-wire memories have to be read in a way that inhibits their use as random-access devices. Usually you have to read everything from the beginning to a specific memory location, and that is not very handy except for program storage and databases that may be read in "all at one time."

There are two ways of organizing a parallel processing instruction set that executes a program:

- Write one master program that is intelligent enough to sort out the needs of the overall process and then assign as many cogs as are necessary to the various tasks that have to be undertaken automatically. At this time, the software (meaning the sophistication) to do this does not exist within the Propeller/Spin system.
- Let a human programmer take a close look at the program requirements and decide what each of the cogs will be assigned to do. This is the way we will undertake our tasks because this is what the Spin language is designed to do. This means that we have to have a good understanding of exactly what needs to be done to solve the

problem under consideration. Although this may seem trivial at first, it is a relatively sophisticated undertaking complicated by the fact that the solution is not unique. When there are many ways to solve a problem and the best way is not always apparent, things can get complicated. It takes a long time and lots of experience to get good at solving problems in a parallel environment. You have to develop a certain amount of "expertise." Even so, we will come up with some basic guidelines about how to proceed as we learn more about the cogs and how to use them in their parallel environment.

Static Versus Dynamic

Basically, the types of tasks we are interested in can be divided into two major categories: those tasks that do not change as we process the data and those that do. According to the classifications I will define, tasks involving data that does not change are classified as *static* tasks and those handling data that changes in real time are classified as *dynamic*.

STATIC SYSTEMS

A large number of tasks handle large amounts of data that either does not change or changes only marginally over short periods of time. An example of such a task would be a list of customers an insurance company maintains. We may need to sort this list from time to time, in any number of different ways, to get information about its contents for our current needs, but there is no need to respond to something critical in real time. For example, we might be looking for everyone over 90 years of age to send them a wellness greeting, and if we can send the information out by the end of the day it will in most probability be okay. Here are some other examples of static systems:

- Mailing lists
- Census data
- Payrolls
- Calendars

In general, these databases do not change over short periods of time and their basic use involves the data itself rather than something dynamic that is happening within the system. You may be interested in the demographics of the U.S. population in 1873 and there is not much that is going to change within the database that represents that population in the next few days (other than newly discovered historical data that may be added from time to time).

Handling static tasks requires speed but not necessarily parallel processing, although we had agreed earlier that almost every task could be done faster in a parallel environment. Even speed is not paramount in that a large mailing list can be sorted overnight, and for most purposes that would be acceptable.

Parallel processing is not particularly well suited to handling such static systems, although as time goes on, and massively parallel systems become available in both hardware and software, parallel engines will be used to handle the kind of databases we have been discussing.

DYNAMIC SYSTEMS

On the other hand, a number of other systems contain information that is changing constantly, and in many cases we need to read and manage the various properties of such a system in real time so that we can control the system to get the results we have specified. Almost every industrial control situation is a dynamic system with more than one variable. By definition, in any dynamic system things change constantly.

One aspect of interacting with a dynamic system is looking at (reading) the variables that are changing. We are interested in these variables because they define the operation of the system. In most cases, the data of interest is represented by the many displays on the control panels that manage what is going on in the system. These displays are designed for human observation and response. In most cases, the management of the system itself is undertaken by some type of automatic mechanism executing a complicated algorithm. Each control loop has its own feedback system and has to be managed by some sort of machine intelligence. Today, this intelligence is provided by computers, both small and large. Such systems are particularly well suited for management by parallel systems where any number of variables may affect one outcome. Complex formulas, fuzzy logic, and machine intelligence play their role in controlling these systems. (In the old days, complicated systems were handled by experienced operators who had long-term experience with the systems. The operators were accepted as being experts in their fields.)

An example of a simple dynamic control system is a domestic hot water heater. The controller turns the heat on at a certain temperature and turns it off at a higher temperature. If the temperature gets too high, another control might override the heat input or a relief valve might be used to release the built-up energy in a safe way. We do not need a sophisticated control system here. The available systems are reliable, durable, and safe, and the system response is fairly slow.

On the other hand, an automated baking line in a modern bakery needs a very sophisticated control system, which has to be manipulated constantly to get the perfect cookies we expect on our grocery shelves. Machine intelligence and fuzzy logic have application here. The inputs of such a system include the following:

- The temperature in the oven
- The speed of the conveyors in the oven
- The temperature history for the last hour along the long linear oven
- The ambient external temperature
- The humidity at the oven door
- The color and variety of the flour being used and the year the grain was harvested in
- The color of the finished product, top and bottom
- The results of the last batch
- Conditions halfway down the baking oven

In most of the situations we would consider, most of the parameters would affect one another. They would be described as being interactive. Some would provide positive feedback, and some negative. In such situations, the management of the controlling functions can get quite complicated. Parallel processing is exquisitely suited to such control situations. Of course, as beginners we are not about to start programming the system for a modern factory with 2,500 interactive feedback loops, but the simple systems we will be looking at do represent the kind of situations we can expect to see in our workplaces on a regular basis. Understating these simple systems will prepare us to understand more complicated systems down the road.

ANOTHER, SIMPLER EXAMPLE

The running of a stepper motor represents a simple-but-difficult-to-manage system that needs constant attention in real time if we are to attain and maintain the high speeds that are often needed. The problem is especially interesting because there is nothing about the system that we do not understand, meaning that we know exactly what has to be done. The question is, *how* do we get it done? The running of a bipolar stepper motor is covered in detail in Part III of this book.

In most of the applications we have in mind for our simple parallel systems, we need to display, read, set, and manipulate any number of variables. We will assign the cogs to tasks such as the following:

- Managing the LCD display
- Reading one or two potentiometers for input
- Making simple calculations and comparisons
- Reading information from various system components
- Setting high and low limits for the controllers
- Reading a keyboard
- Communicating with a computer
- Managing and annunciating alarms

As a general rule, in our experiments we will always use the LCD both to show us what is going within the system and to display the results of our efforts. We always read one or two potentiometers to provide the input variables we need. We will use one cog just to manage the device we are experimenting with, and other cogs will be added as needed.

One Cog

The preceding being the case, we need a solid understanding of what each cog is capable of doing. The eight cogs are identical in every way, and if we understand one cog, we will understand them all.

Each cog has the following exclusive and common (shared) memory resources available to it:

- Two kilobytes of RAM organized as 512 32-bit longs
- Access to 32KB of common RAM, addressable as bytes, words, or longs
- Access to 32KB of common ROM
- Access to clocks, locks, and other devices that make the system work
- Access to a system counter

Individually, each of the cogs in turn has the following capabilities:

- Counter A
- Counter B

The counters can each execute a task independent of all other processes. A total of 16 counters are available for this. (Two each in eight cogs.)

- A video generator

Other features include the following:

- Mutual access to certain memories
- Exclusive access to certain memories
- Access to the 32 I/O lines by all the cogs at all times
- Settable oscillator/clock speeds
- I/O output register
- I/O direction register

I/O input register can be read to determine the current state of all I/O pins.

The external I2C EEPROM mentioned in the first paragraph of the previous chapter is not available to individual cogs. It downloads into hub memory (the main memory), and only at startup. This happens only if no PC is attached to the Propeller. If a PC is present, the system looks to the PC to receive its program.

When downloading a program from a PC to the Propeller system, you have the choice of downloading the program to the attached external EEPROM or to the on-chip main hub RAM in the system. The program and subprograms you create are targeted to available cogs, one after another. Because all cogs are identical, it does not matter what code is assigned to which cog. The programs in the various cogs can interact with one another via the shared memory. The system is completely flexible—which unfortunately means there are a lot of ways to get it wrong. It takes a formal and disciplined approach to get everything right. In this book, we will develop some of the techniques needed to assign cogs their responsibilities in an orderly way.

The system will operate as follows for the programs we will write:

The first cog will control the system on initial startup. It runs the initial program and assigns the execution of the various subprograms (that are part of the main program) to the other cogs as specified in the main program. The Spin Interpreter is copied from main ROM to cog RAM for each cog that will execute Spin code. The hub assigns system control to each cog, in a round-robin fashion. At this stage, we do not need to understand the details of how all this happens, but we do need to understand that this overall process exists.

- Once the system is up and running, all the cogs are equal and any one of them can stop or start any other cog and assign it a program to run. All cogs can modify the clock speed. They are equal in every way.

The Propeller Tool allows you to write programs on a PC and move them to and execute them on a Propeller chip with one keystroke. Using the tool is easy and intuitive. The difficult part is learning to use the Spin language. The Propeller can also be programmed in the Assembly language (Propeller Assembly) provided by Parallax and described in the Propeller Manual, but we will not cover that language in this book. (In addition, third-party C compilers are available for the Propeller chip, but using them is outside the scope of this book.) Everything can be done in Spin except tasks that require extreme speed.

The use of the Propeller screen is discussed in detail in the Propeller Manual (see Figure 7-1). It is well worth doing all the exercises provided therein.

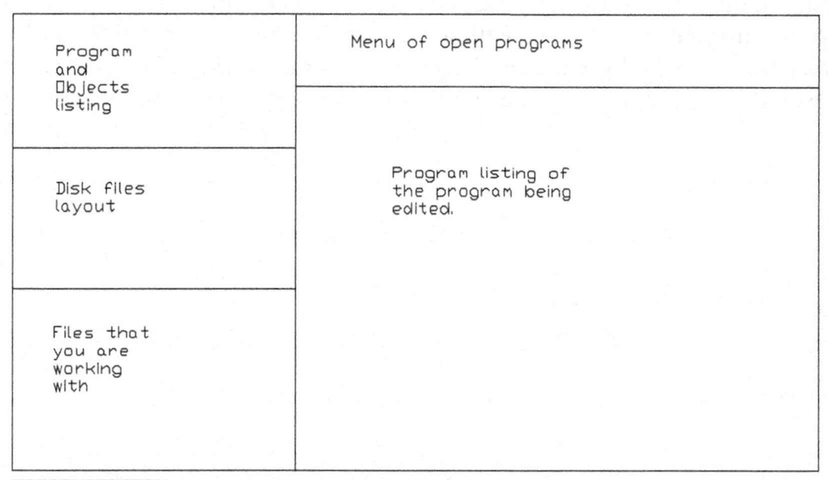

Figure 7-1 The Propeller Tool screen layout

Counters

Note *Counter modules (sometimes referred to in this text as "counters") are an advanced subject not suitable for beginners. We are going to need to generate a pulse width modulated (PWM) signal for many of our experiments, so we need to talk about just this one application of a counter in a very cursory manner. Do not worry if you do not understand this in every detail just yet.*

Each cog contains two identical, independent counter modules that can be configured to perform a variety of tasks—some independent and some cooperatively with the cog. These counter modules are typically referred to as Counter A and Counter B, and on startup they are both disabled (off). Because the counter modules are capable of operating independent of the cog, they have no effect on the cog's execution speed. Although each counter module has 32 different modes of operation and a myriad of applications, in this beginner's book we will only discuss one use of these counter modules: pulse generation.

It is important to emphasize that the operation of the counters can be independent of the operation of the cog once they have been set up. The cog can continue to fetch and execute Spin tokens while the counters do whatever they get configured to without interference between the two. Another approach for some applications involves code that modifies and/or reads the counter module's registers. Depending on the application, this can be done once or periodically.

In this next experiment, we will use one counter module to generate a series of pulses called a PWM (pulse width modulated) signal. The PWM signal's pulses will repeat at a regular interval, as shown in Figure 7-2. This interval is called the signal's period or cycle time, and repetitions of the signals are sometimes referred to as cycles. The duration of the pulse during a given cycle can be varied. These variations can be used for information exchange with devices such as servo controllers and TV remotes or for motor control by limiting the amount of time during each cycle that current is allowed through a transistor that supplies the motor. We will use a counter module to

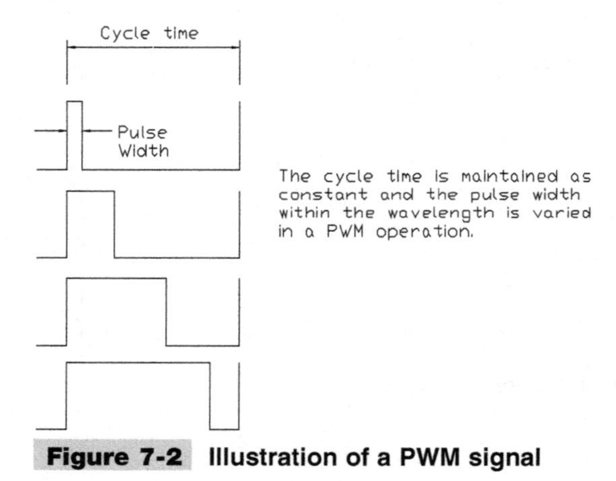

Figure 7-2 Illustration of a PWM signal

vary the pulse durations because it is a very precise tool for this task. Because the counter module can be configured to deliver the pulse independently, this makes it possible for the cog to attend to other tasks while the pulse gets delivered.

Counter: General Description

In this experiment, we discuss Counter A (CTRA); however, counter B is identical. For a counter module tutorial, consult the "Counter Modules and Circuit Applications" chapter in *Propeller Education Kit Labs: Fundamentals.* Also, read the descriptions of the counters in the Propeller Manual. (That material is not repeated here.) Read the sections on the CTRA, FRQA, and PHSA registers. The three registers work in concert, so their functions and interactions need to be considered collectively.

Each counter is configured by a 32-bit control register. Counter A is configured by the CTRA register, and Counter B is configured by CTRB. Each control register contains bit fields with the following information:

Specification of a counter mode	CTRMODE
Identification of pin A	APIN
Identification of pin B	BPIN
Specification of a dividing factor	PLLDIV (Div factor is for PLL mode.)

The information is entered into the control register by setting and clearing relevant bits in the register. We have control of all the bits in the register, but not all bits are relevant.

The counter "mode" that we specify determines how the counter will behave. In the PLL mode, the dividing factor manages how rapidly the process progresses. All this will be much easier to understand as we look at the PWM example identified earlier.

Assignment of the 32 Bits in Each of the Counters

Of special interest to us at this time is the specification of the pin number that is going to output information from the counter. This number, which can vary from 0 to 31, is specified in bits 0 to 5 and is called the APIN bit field. We will use I/O pin P7. If the counter gets configured to interact with a second I/O pin, that number would be specified in bits 9 to 14, which is the BPIN bit field. Our example will only use the APIN field. If we were to use the phase locked loop (PLL) modes for generating high frequency signals, we would also need to specify a number in the PLLDIV bit field (bits 25..23) as a step in determining the final frequency. We will not utilize PLL mode in this text, so the details are not covered here. However, you can find out more about it in the Propeller Manual as well as in the *Propeller Education Kit Labs: Fundamentals* textbook. Both are available for download from Parallax.com.

The 32 modes in which the counter can operate and the use of these modes are beyond our interests at this time; however, the 32 modes are specified by five bits in the counter control register. These bits go from bit 30 to bit 26. We are interested in a PWM operation. This is specified by setting these bits to 00100. See Table 2-7 in the Propeller Manual. An application note that describes counter operations is available from the Parallax downloads site, but it is rather advanced for beginners. (The first mode, 00000, is "counter off.")

Using Counter A for PWM Generation

A close look at Figure 7-2 indicates that we have to coordinate the operation of two signals. We have to make a signal go high at a fixed rate to set the cycle time, and we have to maintain a variable cycle time within this signal to set the pulse width. (Counters are much more complicated and versatile than that, but this is enough information for us at this time.)

First, we have to specify the mode we want the counter to operate in. The single-ended PWM mode is specified by setting CTRA[30..26] to binary 00100 with the following code:

```
ctra[30..26]:=%00100
```

Next we identify pin A as the pin the signal will be expressed on:

```
ctra[5..0]:=Pin
```

We set the number that will be added (or subtracted if negative) to the counter during each cycle as follows:

```
frqa:=1
```

The counter is now set up. Next, we have to start using the counter. To make sure nothing we did not plan for will happen, we make the system dormant by setting the starting pulse width to zero:

```
PulseWidth:=0
```

The cycle time is set by dividing the clock frequency in CPS by the frequency. The clock frequency is 10_000_000 CPS. We want a cycle time of 10 milliseconds. This specifies a frequency of 100 cycles per second, so we use the following:

```
Cycle_time:=clkfreq/100
```

We need a starting point for the counter because we will be using the system counter to measure various cycle times and pulse widths. We do this by setting the starting period count because we will set the final period count by adding the current counter value to the cycle time:

```
period:=cnt
```

With all this in place, the code in Program 7-1 will give us a fixed pulse width of PulseWidth. The pulse width is set as a negative value because we will be adding 1 from the counter at each cycle of the clock to work it to zero.

Program 7-1 Segment: PWM Routine

```
PRI
repeat                          'power PWM routine.
   phsa:=PulseWidth             'Send a high pulse for PulseWidth counts
   period:=period + Cycle_time      'Calculate cycle time
   waitcnt(period)                  'Wait for the cycle time
```

Program 7-2 allows us to look at the result with an oscilloscope. We will use a fixed pulse width that is half the cycle time (50_000 cycles) so that our success will be readily verifiable on the oscilloscope.

Program 7-2 Generating a Fixed PWM Signal

```
{{14 Sep 09   Harprit Sandhu
PWM.spin
Propeller Tool Ver 1.2.6
Chapter 7 Program 2

This program generates a fixed PWM. Meaning that as programmed
the width does not vary. (Set at 50%) It is easy to recognize
}}

CON
  _CLKMODE=XTAL1+ PLL2X           'The system clock spec
  _XINFREQ = 5_000_000            'crystal frequency

VAR
  long pulsewidth
  long cycle_time
  long period

PUB Go
  dira[7]~~                        'set output line
  ctra[30..26]:=%00100             'run PWM mode
  ctra[5..0]:=7                    'Set the "A pin" of this cog
```

(continued)

Program 7-2 Generating a Fixed PWM Signal (*continued*)

```
frqa:=1                              'Set this counter's frqa value to 1
PulseWidth:=-5000                    'Start with position=5_000
Cycle_time:=clkfreq/1000         'Set the time for the pulse width to 1 ms
period:=cnt                      'Store the current value of the counter
repeat                           'PWM routine.
  phsa:=PulseWidth               'Send a high pulse for PulseWidth counts
  period:=period + Cycle_time    'Calculate cycle time
  waitcnt(period)                'Wait for the cycle time
```

Run this program and look at pin 7 with your oscilloscope. You should see a signal with a 50% duty cycle. (Now if we can somehow vary the pulse width, we will have a PWM generator.)

Accordingly, we address varying the duty cycle of the PWM signal from 0% to 100%. We will do this by reading a potentiometer and using its input to vary the pulse width as needed. We need to make some modifications to the program to do this. Because you are not yet proficient enough to do this, I will just give you the code we need in Program 7-3. In order to run this program, you need to load the program Utilities and the program LCDRoutines4 to the same file as this program to make it all work. (They actually have to be saved to disk in the same file as your program for your program to be able to find them.) We require these objects because we need to call certain methods that are resident in these objects.

Program 7-3 Variable PWM Signal Based on a Potentiometer Reading

```
{{14 Sep 09   Harprit Sandhu
PWM1.spin
Propeller Tool Ver 1.2.6
Chapter 7 Program 3

This program generates a variable PWM signal.
Based on a potentiometer reading
}}

CON
  _CLKMODE=XTAL1+ PLL2X             'The system clock spec
  _XINFREQ = 5_000_000              'crystal frequency

VAR
  long stack1[25]                   'space for motor
  long stack2[25]                   'space for LCD
  long pulsewidth                   '
  word pot                          '
OBJ
  LCD     : "LCDRoutines4"          'for the LCD methods
  UTIL    : "Utilities"             'for general methods
```

(*continued*)

Program 7-3 Variable PWM Signal Based on a Potentiometer Reading (*continued*)

```
PUB Go
  cognew(RunMotor(7),@Stack1)
  cognew(LCD_manager,@stack2)
  repeat
     Pot:=UTIL.read3202_0                'read the pot at MCP3202  line 0

PUB RunMotor(Pin)|Cycle_time,period     'method to toggle the output
  dira[7]~~                    'gain access to these three amplifier lines
  dira[19..20]~                'potentiometer location
  ctra[30..26]:=%00100         'Set this cog's "A Counter" to run PWM
  ctra[5..0]:=Pin              'Set the "A pin" of this cog to Pin
  frqa:=1                      'Set this counter's frqa value to 1
  PulseWidth:=50               'Start with position=50
  Cycle_time:=clkfreq/1000     'Set the time for the pulse width to 10 ms
  period:=cnt                  'Store the current value of the counter
  repeat                       'power PWM routine.
    phsa:=-(pot*244/100)       'Send a high pulse for PulseWidth counts
    period:=period+Cycle_time  'Calculate cycle time
    waitcnt(period)            'Wait for the cycle time to complete
PRI LCD_manager
  LCD.INITIALIZE_LCD                    'initialize the LCD
  repeat                                'LCD loop
    LCD.POSITION (1,1)                  'Go to 1st line 1st space
    LCD.PRINT(STRING("Pot=" ))          'Potentiometer position ID
    LCD.PRINT_DEC(pot)                  'print the pot reading
    LCD.SPACE(5)                        'erase over old data
```

Run the program and turn the potentiometer end to end. You should see a fixed cycle time in which the high portion of the wave goes from 0% to 100%.

8

THE EIGHT COGS

The eight cogs are identical in every detail. Therefore, now that we have a feel for what one cog can do, we need a general overview of the eight-cog environment. This chapter considers the eight cogs as a whole, along with their ancillary control hardware. You should be completely comfortable with the previous chapter before you start this chapter. You need to understand what the capabilities of one cog are to understand them altogether.

The Cogs

Each of the 32-bit processors in the Propeller chip is called a cog. The Propeller chip has eight of these cogs on it. The eight cogs can be run simultaneously, and each of them can perform one or more tasks, depending on how the tasks are designed and programmed. A cog can be programmed to perform more than one task if the tasks can be performed in the time available and do not interfere with one another. Because the Propeller system does not support the use of interrupts, anything that would have needed an interrupt in the usual single-processor environment needs to be assigned to its own cog in the parallel Propeller environment. There is no easy way to avoid this requirement, and because you have eight cogs at your disposal, there is usually no need to.

Each of the cogs is a 32-bit logic engine with a sophisticated instruction set. Between them, the eight processors have access to 32KB of hub RAM. Each of these cogs can access this memory, one at a time, in a round-robin fashion. The access to the memory is synchronized by the system clock and is controlled by a hardware device called the system hub. Each cog gets access to the shared memory for the same amount of time.

All the cogs can access the 32 input/output lines simultaneously. The timing constraints of how these interactions take place and what the latency is (meaning the worst-case delay) are described in the owner's manual and should not be a concern at this stage of the learning process. Timing critical tasks is beyond what we need to know as beginners.

There is no user-accessible stack (for subroutine return addresses, and such) either common or specific to a cog, and there is no interrupt function anywhere within the entire system. Neither of these functions is needed because having eight cogs running in parallel essentially eliminates the need for them.

By far the greatest shortcoming of the system is the lack of large amounts of memory. Not having a large amount of memory eliminates the possibility of undertaking large number-crunching operations. The chip is not designed for large number-crunching operations, but as a general-purpose controller it has more than adequate memory. Serial memory can be added to the system, but serial memory is not as fast as RAM and has read/write issues related to accessing it regarding the speed and rules of the operation. If a serial one-wire memory has to be accessed repeatedly, a lot of time is used up. Graphics applications that use large displays or run simulations also need large amounts of memory and therefore have the same problems. At our level of interest, the memory is not a concern. We have more than enough. None of the programs in this book come close to needing but a small part of the memory at our disposal.

The Flags

The system provides eight flags (which are referred to as "semaphores" or "locks" in the literature) to allow various processors to access resources that are shared in an orderly manner. This is done by setting and clearing the semaphores. The semaphore system is not a concern for beginners, but we should have an awareness of the existence of these flags so that when we come up against these difficulties, we will know that the solutions have already been implemented by the machine designers.

Special Memory Locations

The shared memory consists of 32KB of hub memory organized as 512 longs. Part of this memory is reserved for special-purpose registers that determine internal cog relationships. These are listed in the owner's manual.

The System Clock

The system clock coordinates the operation of all the hardware items in an orderly fashion. The system clock rate can be controlled by the Propeller chip (although none of the programs in this book use that capability). The resister/capacitor combinations internal to the system allow the system clock rate to be used without an external crystal. It is also possible to use an external crystal to set the clock rate and use a multiplier to increase it. The decisions as to which system will be used and the frequency at which the system will run are made by the programmer and the hardware designer, depending on how the system is to be used. For most of our experiments, we will use the 5 MHz crystal that comes with the education kit. We will use a multiplier of 2 for an effective rate of 10 MHz for the clock. We specify the operational frequency at the top of the program under the Constants assignment.

As beginners, we do not need to be concerned about the finer details of how the system clock operates. Leave it at 10 MHz for all the experiments for now.

Programming

The program is written with the Propeller Tool, which provides a complete programming environment. If you prefer, the program can be written in Word or any other word processing program and then cut and paste it into the Propeller Tool. If you want to have circuitry as a part of the program documentation, the circuitry must be created in the Propeller Tool environment with the Parallax font.

Once written, the program can be transferred either to RAM or EEPROM and executed from either one. In either case, the program has to be loaded into the main memory of the Propeller under consideration before it can be executed. If the program is in Assembly code, that code will be executed as such. If the program is written in Spin, it will be interpreted by the Spin Interpreter during run time and executed.

The ROM

The read-only memory contains the Spin Interpreter, math tables for generating all the related mathematical functions, sin and log tables for making mathematical calculations, and font descriptions for the font used by the system.

For now, we do not need to worry about how all this takes place because it happens automatically, in the background, when we are using the Spin language.

9

SPECIAL TERMS AND IDEAS

This chapter introduces you to several new terms and ideas used by the parallel-processing discipline that cover the new concepts, software, and hardware created for working in a parallel-processing environment.

The ideas we are interested in are expressed in the Spin language environment, as implemented by Parallax, to allow us to program the Propeller chip. The Parallax software engineers have decided to design the Spin language in a way that you will find both easy to use and powerful once you get used to its structure. They also decided to coin a few terms that allow us to differentiate discussions in and about this language to provide a certain amount of semantic isolation.

For most of us, parallel processing is a new concept. Even fairly competent programmers need to rethink how they approach a problem before they start to program in a parallel environment. The major concept that has to be addressed is assigning the various parallel tasks to the eight processors in an intelligent fashion. This can be done in two ways. In the first, we write a program just as we would for a one-processor linear system, and the computer is so smart that it figures what has to be done and how it is to be done in a multiprocessor system. The second method involves the programmer breaking the task down into well-thought-out subtasks and then assigning the subtasks to the eight processors, as appropriate. Because each cog is capable of doing more than one thing, the problem lends itself to elegant and innovative solutions. As things stand as of this writing, the artificial intelligence needed to allow the computer to assign the tasks autonomously does not exist for the Propeller. We have to resign ourselves to assigning the tasks to the various subprocessors individually as we design the application.

The Hardware

The hardware is already set to be used as a parallel environment, with eight cogs, 32 input/output lines, and shared memory. This is set in concrete. Nothing that we can change is available to us on the hardware side. What we do have available to us is the freedom to use the hardware as we see fit. Although this may seem trivial at first, as we will see, it is not. There are intelligent ways to use the hardware—and there is much room for foolishness.

The Software

The software introduces us to some new concepts and constructions to allow us to play in a parallel environment; therefore, understanding the software is the key. It's not just understanding what each instruction does; it is understanding what the motivation is behind designing the software the way it was designed so that our efforts are not contrary to the way the system is intended to be used. In other words, we do not want to row upstream. Even so, each programmer will have his or her own way of doing things. Each project undertaken has to provide a satisfactory solution, and there are many ways to get the job accomplished. The fact that we have eight processors at our disposal makes for many, many interesting possibilities.

As we look closely at the system, we will see that some of the realities are forced upon us by the features selected by the hardware designers and some by the choices made for us by the designers of the software. Because this was all decided some time ago, we have no say in the matter. We have to work within what has been provided to us.

Here are two basic concepts that need to be understood:

- If a certain item of hardware does not exist, there is neither a need nor a way to address it. It needs no software support.
- If we have some great piece of hardware and the software to address it does not exist, there is no way to give the hardware the instructions needed to manipulate its properties. We can only ignore the hardware.
  In other words, if there are only 32 I/O lines on the processor, no amount of software sophistication will allow us to turn I/O line 33 on or off, and if the hardware has 33 input/output lines and the software allows us to address only 32 of them, the 33rd line is useless for all practical purposes.

In philosophical terms, we say that we cannot discuss those things for which there are no words. First, the words need to be invented and defined. The Spin language does this for us to allow us to proceed with programming. That is what the instruction set of every language does.

New Hardware-Related Definitions

On the hardware side, the new definitions are as follows:

- **Cog** A cog is an independent 32-bit RISC-like processor that resides within the Propeller chip. The program that the cog executes is downloaded into Cog_0, along with all other program instructions. The interpreter is transferred from main ROM to the cog when each cog is opened. An area of memory has to be assigned to every cog except Cog_0 in the main memory for its program space. Cogs are given access to shared resources, one after another, in a round-robin fashion by the hub.

- **Hub** The hub is the central monitoring device that controls which cog gets to do what and when. It allows access to the shared system resources by each cog, one at a time, in a round-robin fashion. It manages all ancillary functions for the cogs. Each cog is assigned the same amount of time during its turn to access the mutually exclusive resources.
- **Shared memory** This is the internal memory (to the Propeller) that all the cogs can access when it is their turn to control the system. It is the hub RAM. All the variables declared in the VAR section of the program are stored in the shared memory, regardless of whether more than one cog addresses them. The shared memory is not to be confused with the external memory that the program is read from on startup if no PC is attached to the Propeller system. The external memory is not accessed by any cog directly, although it does download to the Cog_0 memory space on initial startup and is loaded through Cog_0.
- **System clock** Like all system clocks, the Propeller system clock times all internal operations. There is only one system clock, and it is shared by and accessible to all the cogs. Its speed is programmable.
- **Round-robin** This refers to the serial access the hub provides for all exclusive resources to each of the cogs in the system. Each cog gets the same amount of time as every other cog. Turning off a cog saves energy but does not save time. In other words, turning off a cog does not speed the system up.
- **External memory** This is the "one-wire" memory that is external to the Propeller chip. This memory contains the program the Propeller will execute if it is not connected to a computer on startup. This means you cannot have a free-standing device if it does not have external memory as a part of the Propeller's peripherals.

New Software-Related Definitions

On the software side, the new terms and definitions are as follows:

- **Object** An object is a piece of software that contains any number of methods that can be called by other programs if they have been defined as PUB or public routines (methods). Any program that wants to use the programs in an object has to declare its intentions in the OBJ part of the program. An object itself does not have to be executable, although most are.
- **Methods and nesting** An important part of the way the Spin language is defined is the lack of a RETURN statement and the way in which this fact is compensated for. In the Spin language, a subroutine is called a "method." All methods are terminated with a blank line. All statements that are a part of a repeat structure in a program listing are made part of the repeat structure by insetting them by one or more spaces. This is illustrated in Program 9-1.

Program 9-1 Segment Illustrating Indented Lines

```
repeat                                   'movement loop
  iters:=pot2                            'set number of iterations to perform
  index:=0                               'reset index
  repeat iters                           'do iterations
    index:=index+1                       'increment index
    targetPosition:=targetPosition+index    'set new position
    waitcnt(clkfreq/Pot1+cnt)               'wait time for iteration
  repeat iters                           'now do the slow down
    index:=index-1
    targetPosition:=targetPosition+index
    waitcnt(clkfreq/Pot1+cnt)
  repeat while startFlag==0              'wait till done
  waitcnt(clkfreq+cnt)                   'delay to see stop
  targetPosition:=startPosition          'set to go back
  waitcnt(24_000+cnt)                    'wait to get done
  repeat while startFlag==0              'wait till done
  waitcnt(clkfreq+cnt)                   'delay to see stop
```

- **Process** A program is called a "process" in the Spin language.
- **Method** Subroutines are called "methods" in the Spin language.
- **Latency** Latency is the time between when an action is required to take place and when it actually takes place. The concept is particularly relevant in the Spin language because each cog gets access to the mutually exclusive resources only when it is that cog's turn. So the possibility exists that when a cog is about to ask for some piece of information, it becomes some other cog's turn and the first cog now has to wait until its next turn to get the information it requires. This is the worst-case latency for this particular scenario. See the Propeller Manual for a discussion of this topic.

  Other processors also have latency, but this is not as critical as it is in the Propeller system because of the multiple processors and the resource sharing requirements.
- **Variable assignments** Variables are assigned in two ways: either as shared variables or as variables local to a method. Shared variables are declared under the VAR block, like this:

```
VAR
  long     Pos[3]           'Create buffer for encoder
  long stack2[25]           'space for Cog_LCD
  long stack3[25]           'space for Cog_SetMotorPower
  long stack4[25]           'space for Cog_RunMotor
  long stack5[25]           'space for Cog_FigureGain
  long stack6[25]           'space for Cog_Start
  long stack7[25]           'space for readopts
  long pulswidth            '
  long startPosition        '
  long PresentPosition      '
```

```
    long TargetPosition      '
    long PositionError       '
    word startFlag           '
    long gain                '
    long pot1
    long Pot2
    word iters               'number of iterations
    word index
```

Local variables are declared on the first line of the method, like so:

```
PUB POSITION (LINE_NUMBER, HOR_POSITION)|CHAR_LOCATION 'Position the cursor
'Line Number : 1 to 2
  'Horizontal Position : 1 to 20      'specified by the two numbers
  CHAR_LOCATION :=(LINE_NUMBER-1)*64 'figure location. See Hitachi HD44780
  CHAR_LOCATION +=(HOR_POSITION-1)+128 'fig loc. See Hitachi HD44780
  SEND_INSTRUCTION2 (CHAR_LOCATION) 'send the instr to set cursor
```

In the preceding code, LINE_NUMBER, HOR_POSITION, and CHAR_LOCATION are local variables available only in the public POSITION method. Local variables are stored in the main memory, not in the cog memory. They do not take up cog memory space.

10

THE SPIN LANGUAGE

Note *I find that the best way to use the Propeller Manual is to have a hard copy in hand for reference and the latest version of the manual in electronic format open in a window on the computer screen. This way, you can search as well as cut and paste from the manual rapidly when you need to. However, nothing beats having the book in your hand—and nothing beats being able to read in bed!*

Note *In the program listings, I have tried to use a straightforward programming style that's easy for beginners to read and understand. If you look at the published listings, you will see that it is possible to write much more efficient code, but this would make the code harder to read. I have tried to avoid this whenever possible. Once you get comfortable with Spin, you can convert what I have provided to a more efficient format that suits your programming style.*

You should have a hard copy of the Propeller Manual (PM) in your possession for reference as we proceed with this chapter. This is especially important because, as of this writing, the manual is the definitive authority on the Propeller chip, the Spin language, and the Assembly language. You cannot do much without it.

Almost all experiments in this book are performed on and with the Propeller Education Kit as provided by Parallax. I used the version with the 40-pin chip. All illustrations show wiring on this kit. All wiring diagrams are for this kit, although they all will work on other hardware. Both the Development Board and the Demo Board are also suitable, but minor modifications to the software might be required to reflect the appropriate addressing changes you find necessary.

Before we can start writing simple routines that will allow us to learn how to use the Propeller chip, we need to have a rudimentary understanding of how the language is used to write programs. The new higher-level language Parallax has designed specifically for this chip is called Spin. Spin is an object-oriented language that shares some properties with simpler languages such as BASIC. However, it is a very powerful language, and you will find that it is not hard to learn, especially considering that there is no alternative other than the even-harder-to-learn Assembly language (PASM) if you want to use the Propeller chip.

We will not cover the sophisticated intricacies of the language in this book but rather concentrate on using the simplest of the commands in the language to perform the simplest tasks so that we can start to become comfortable with the language. We will move toward using the language in a more sophisticated context as your skills improve and your comfort level with the language increases. In the first few examples, I provide code written in Spin that looks as much like BASIC as I can make it so that you can more easily follow what is going on. For those who are not comfortable with BASIC, I suggest you either get a beginning BASIC text or just ignore the BASIC examples and concentrate on learning what we are trying to do with Spin.

The first program is the ubiquitous "blink one LED" program that seems to be a requirement in most texts. Before we can write this program, we need to say something about how the code is specified in Spin. Spin is *structured,* meaning that it requires you to be more organized in the way you do things. It requires that constants and variables be specified before you start writing the body of your program. It means that routines have to be specified as being private or public, and it means that a program has to follow a certain general structure. It also means a host of other things, which we will discover as we go along.

Spin uses a documentation scheme that allows you to do the following:

- You can embed documentation within the program in a way that allows you to suppress two levels of the documentation.
- You can separate the documentation from the code when needed.

These are extremely useful features, and every attempt should be made to follow the suggestions given in the Propeller Manual in regard to the use of these features.

The Spin screen is color-coded so that each section of code is easy to recognize as a separate block. Although you can specify the color codes, we will not bother with that in this book. We will use the color scheme that comes with the Propeller Tool.

Six separate block designations are used in the Spin language:

- **CON** Used for the specification of constants
- **VAR** Used for the specification of variables
- **OBJ** Used for the designation of objects to be called
- **PUB** Used for declaring methods as public
- **PRI** Used for declaring methods as private
- **DATA** Used for declaring data and data structures

All programming segments must fit within these six blocks. However, comments and documentation can appear outside the blocks. We will cover the use of the blocks in more detail as we need them.

Let's begin with a few instructions that will be used in almost every program we will ever write. The first ones have to do with declaring constants of all kinds.

CON

The CON statement is used to declare the constants that will be used in the program. In any program, the first thing we need to do is to declare and define the constants we are going to use within our program. Defining the constants also names the constants and thus makes it much easier to follow the program. It is best to define each constant on a new line, and every constant within the program should be defined and its use then described adequately. Doing it this way allows you to change a constant at this one location and it will be changed all over the program. This makes it easier to understand and maintain programs.

In our case, we will be using the following two constants:

- **Pin** To identify the pin being used
- **Delay** To specify the delay between the state change of the LED

In the listing, these constants will be typed in as follows (shown in the Parallax font):

```
Pin   =     1     ; number of the pin connected to the LED
Delay =     500   ; delay in milliseconds (half a sec)
```

All the constants declared in the CON section are in the shared memory and are available to all the cogs. Only those constants actually declared in the CON section are available to all the cogs. Constants used by methods that you call from your program are local and are not shared.

The value of a constant never changes within the program, across all methods.

VAR

Variables, of course, are those values that change from time to time within the methods. We define the variables up front so that they are all in one place and are easy to rename and redefine if that becomes necessary. Because Spin uses a number of different types of variables, we have to specify the name of the variable along with the byte, word, or long designation that will be used to contain that variable. In our first program, the variables will all have a value under 255, so we can use the byte variable specification for them. (One byte can hold a value of from 0 to 255 in its 8 bits.)

Putting the variables right under the constants allows us to see all the variables together in one place so that we can add and remove them from the list as necessary. Keeping the variable list next to the constants is an advantage in that if a constant needs to become a variable or a variable needs to become a constant, it can be done from this one place with ease.

All the variables declared in the VAR section are in the shared memory and are available to all the cogs. Only those variables actually declared in the VAR section are available to all the cogs at all times. Variables used in objects that you call from within your program are not shared, and in most cases you may not even know what they are or how they were used in the called method.

OBJ

The OBJ statement defines an object to be called from your program. An object is equivalent to a program and may contain one or more methods. The OBJ section of the program lists the objects that will be called from the program. The objects themselves are defined somewhere else (within limits). The program can call objects or parts of them as needed. It is a good idea to define what each object does in some detail in the documentation so that we know why we are calling it and what it does.

Once these items have been defined, we are ready to start writing the main body of the program. Most programs start with the PUB (public) statement followed by an arbitrary name (it is best to use either GO to indicate the start of a program or MAIN to follow the C convention). In our case, we are going to call all our beginning methods GO. Our first line of executable code will follow the GO assignment.

PUB or PRI

A method can be specified as being public or private. Private methods are available only to the immediate object they are in. Public methods are available to other objects in the immediate vicinity (stored in the same file).

In the Spin language, the specification of every object starts in the first column of text. The body of the object is indented one or more spaces, and everything that is indented under the first line of the object becomes a part of the object or method. There is no "return" at the end of the object, and the object is terminated by the first new block designation encountered. The next unindented line of code after the end of the current object defines the beginning of the next object. So in a way, our main subroutine extends from the declaration of GO as being a public subroutine to the beginning of the next unindented block of code.

Let's get ready to write the pseudo-code instructions that will demonstrate what a Spin program looks like. Here is what we want our first program to do:

- Select and set the LED line to be an output.
- Declare the number of the line to which our LED is connected.
- Turn the LED line to High or On.
- Pause for a short time.

- Turn the LED to Low or Off.
- Pause for a short time.
- Repeat the last four lines.
- End the program.

In order to do all this, a certain amount of housekeeping is necessary before we start. The following minimal setup is necessary for our particular program:

- We need to decide which of the 32 lines available in the Propeller system will be used to control the LED. We will use line 27 so we can save all the pins on one side of the Propeller, from 0 to 15, for future experiments.
- We need to set the selected line to be an output line.
- We need to decide how long the delays will be. We will use 0.5 seconds.
- We need to decide and specify the number of cycles to be completed. We will do it forever.

Note *The cnt variable in Program 10-1 is the system counter in the Propeller. See the Propeller Manual (PM).*

Instead of writing an absolutely minimal four- or five-line program, we will write something a little more complicated and useful. We will still blink the LED, but we will use predefined constants and methods to get a little further along into the Spin language right away. Program 10-1 contains detailed comments on every line and additional general comments to explain what is going on in the program.

We have established a standard that requires the beginning of each program to be a GO method, not unlike using MAIN in the C language. We will follow this standard for all our programs. The simple program structure is illustrated in Figure 10-1.

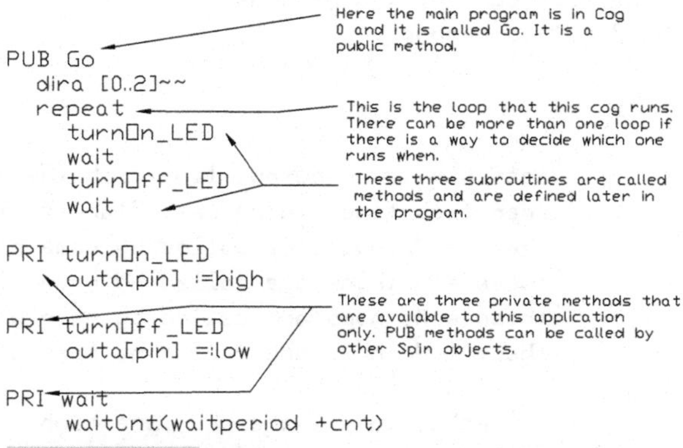

Figure 10-1 **Simple program with three method calls**

All programs are available for copying from the support website.

Program 10-1 Blink an LED Using Methods

```
{{04 June 09   Harprit Sandhu
BlinkMethods.spin
Propeller Tool Ver. 1.2.6
Chapter 10 Program 1

Blinking an LED
This program turns an LED ON and OFF and demonstrates the use
of subroutines in an absolutely minimal way.

Define the constants we will use.

Propeller font schematic:

          |  100 Ω
        21├───WWW───►├──────┐
          |          LED    |
                            ⏚ GND
}}
CON                                  'CON defines the constants
  _CLKMODE=XTAL1+ PLL2X              'The system clock spec
  _XINFREQ = 5_000_000               'external crystal

  pin            =21                  'select the pin to be used for the LED
  waitPeriod     =500                 'set the wait period
  high           =1                   'define the High state
  low            =0                   'define the Low state

{{
The following PUB Go is the main part of the program.
Everything else is in the 3 called methods.
}}

PUB Go
  dira [pin]~~                     'sets pin to an output line with the ~~
  repeat                           'specifies times to repeat. Blank=forever
    turnOn_LED                     'these 4 Methods are called by name alone
    wait                           'these 4 Methods are called by name alone
    turnOff_LED                    'these 4 Methods are called by name alone
    wait                           'these 4 Methods are called by name alone

PRI turnOn_LED                        'Method to set the LED line high
    outa[pin] :=high                  'line that actually sets the LED high
```

(continued)

Program 10-1 Blink an LED Using Methods (*continued*)

```
PRI turnOff_LED                        'Method to set the LED line low
    outa[pin] :=low                    'line that actually sets the LED low

PRI wait                               'Method defines the delay
    waitCnt((Clkfreq/1000)*waitperiod +cnt)   'wait till counter
                                              'reaches this value
```

If there is disagreement between a program in this book and the same program on the website, use the program on the website. It will have the latest changes and error corrections in it. Figure 10-2 illustrates the circuit layout for Program 10-1 on the Propeller Education kit.

You must completely understand each and every line in Program 10-1 before you read another word. This simple program defines the essential structure of almost every program we will investigate in this book, and if you understand it in every detail, you are on your way to becoming a Spin programmer. In later discussions, we will not use words such as "subroutine" but rather will follow the nomenclature used in the Propeller Manual and use "method" so that we can become familiar and comfortable with it. New Spin words will be added to our vocabulary as they are needed for the programs we develop.

Program 10-1 demonstrates the following concepts:

- The CON block is for the constants. In the constants block, we specify what wire/pins on the Propeller are to be used for what function. If we ever decide to attach the LCD to another pin, all we have to do is change the designation in this block.

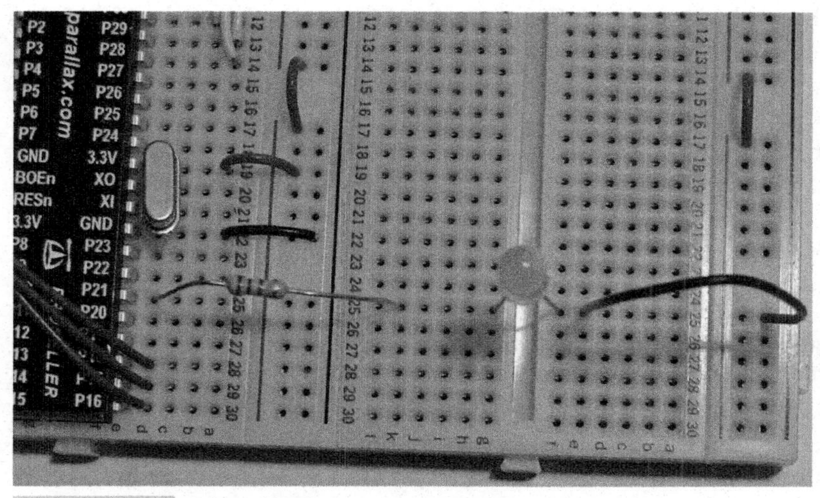

Figure 10-2 Schematic of program with two cogs in it

- The VAR block is for the variables. We will not use any variables in this first program, but in order to demonstrate the use of a VAR block, we must have at least one variable defined in the block. I have defined a dummy variable called "numbr" as a byte. This would mean that "numbr" could have contained any number from 0 to 255 if we were using it in our program.
- The PUB Go statement starts the program proper. The program first defines the pin we are using as being an output and then repeats the turning on and off of the LED an infinite number of times, which is defined by not providing a number after the repeat command.
- The three PRI blocks are methods that are private to this program and are used to define the actions of the three calls in PUB Go. They turn the LED on, create the wait pause, and turn the LED off in the sequence in which they are called.

The purpose of this program is not so much to turn the LED on and off but rather to go through all the procedures we need to go through to write and execute a Spin program. Once you are comfortable with this program, you will be able to write an awful lot of programs in Spin. As we add to the techniques used with the Propeller in subsequent programs, you will acquire more and more of the skills that you need to Spin effectively.

The wiring for Program 10-1 is illustrated in Figure 10-3.

Turning an LED on and off is okay as a start, but we need to input and output all sorts of things into and out of the Propeller if we are to create useful interactions. A most useful addition to any microcontroller project is a small liquid crystal display (LCD). We will implement the rudimentary control of a 16×2 LCD controlled by a Hitachi 44780 controller. Later on, if you like you can extend this implementation to a full and comprehensive control of any display that uses the Hitachi controller. We, on the other hand, will convert the rudimentary LCD display program into code that can be used within all our future programs to display the results of whatever it is that we are doing with the Propeller. This is covered in detail in Chapter 14.

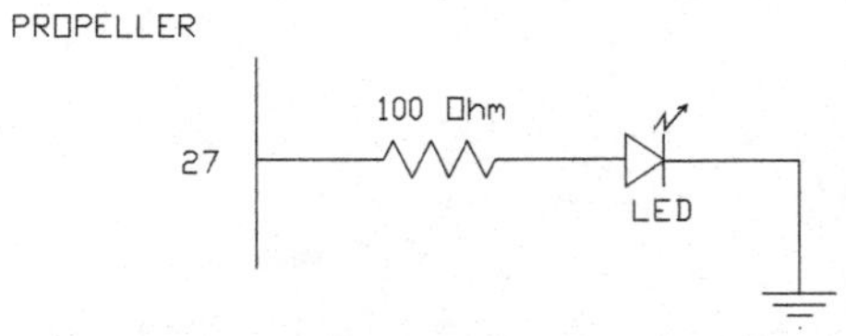

Figure 10-3 Wiring schematic for blinking an LED in Program 10-1

Creating a Program with Two Cogs

Figure 10-4 shows the placement of the various components you need to start a second cog in your system. All other cogs are to be started in this way.

Note *Cog_two will allow us to implement the use of a 16×2 LCD display.*

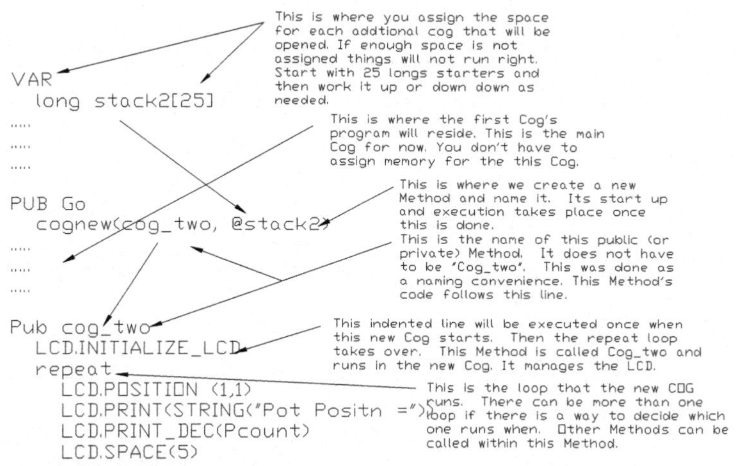

Figure 10-4 Wiring for one LED as programmed in Program 10-1

11

TASKS SUITED TO PARALLEL PROCESSING

Everything (except the most rudimentary tasks) can be handled in a parallel-processing environment. Of course, many tasks are best not handled in this environment. An illustrative example is the display of all the numbers from 1 to 1,000. We could assign this to three processors: one to create all the odd numbers, one to create all the even numbers, and a third to display them all on the computer screen. Obviously, this is not the most intelligent way of getting this job done. Like many other tasks, this task is easily handled by one processor, and that is how we would normally handle it even in a Propeller environment. Essentially, this task is too simple to need a parallel environment.

A parallel environment works best when a number of tasks need to be undertaken simultaneously to complete the job at hand. Complicated-but-not-too-complicated tasks are best picked as our first examples. These and similar examples will be followed up in Part III of the book with full-fledged programming implementations.

Parallel Programming Examples

STEPPER MOTOR

Running a bipolar stepper motor is a good target for parallel programming. The running of bipolar stepper motors is covered in detail in Part III of this book in Chapter 26, and you may want to refer there before you continue reading here. At the least, you should know how a stepper motor works so that what we consider here makes sense to you. Here, we will confine ourselves to the general discussion of the parallel tasks we need to undertake to run the motor.

To run a stepper motor, we need to undertake the following tasks:

- Determine the sequencing requirements for the motor coils.
- Read a potentiometer that we will use to set the speed of the motor.
- Manage the power to the amplifiers for the two motor coils.
- Manage the LCD to display information of interest to us as we run the motor.

Because we are working in a parallel environment, each of these tasks can be assigned to a cog.

DC MOTOR SPEED CONTROL

Controlling the speed of a DC motor with a potentiometer often requires the reading of the potentiometer or some other input device and modulating the power to the motor so that the encoder counts read reflect the desired motor response. In such a situation, a parallel-processing arrangement can be used to our advantage. With the setup we will be using for our experiments, the cogs can be assigned as follows:

- Read the potentiometer that will control the speed.
- Create the PWM signal needed to power the motor.
- Read the encoder repeatedly to get speed feedback.
- Display the results of the experiment on the LCD.

HOBBY SERVO (R/C)

In a hobby servo, the control requirement is to send the servo a pulsed signal every 1/60th of a second. The timing of the pulses is not critical, but the length of the pulses is. They have to be 1,520 microseconds ±750 microseconds long. As usual, we will use the input from a potentiometer to control the length of the pulses. Each critical task will be assigned to a separate cog. With the setup we will be using for our experiments, the cogs can be assigned as follows:

- Read the potentiometer that will control the pulse width.
- Create the pulse width needed to position the motor.
- Send the pulses to the motor at the required times.
- Display the results of the experiment on the LCD.

There are many variations of this control scheme. One obvious possibility is to write the code so that the servo output is a 90-degree quadrant instead of a 180-degree move. In most applications of these servos, the middle 90 degrees of the move is the most useful. It would also be possible to read a second potentiometer and use its value as a trim factor.

SELF-LEVELING TABLE

In a self-leveling table, we need to be able to make a correction both in the X and Y directions. In order to make the correction, we need to detect what the error in the position of each axis is. We get this information from a gravity sensor called the Memsic 2125. We will need a servo for each axis. The two axes will be controlled independent of one another. These functions are implemented by assigning the tasks to the cogs, as follows:

- Read the X and Y errors from the sensor.
- Make an incremental correction to each axis if there is an error.
- Display the status of the table on the LCD.
- Add the reading of two potentiometers to trim the exact level position of the table.

The software developed here could be used to stabilize a camera platform or to provide automatic leveling for a model aircraft.

Summary

In general, every application breaks the tasks to be accomplished into a number of fairly straightforward blocks that are simple to create, and it assigns them to individual cogs. Of course, calculations will also have to be made, and pins will have to be set up to perform the necessary I/O. However, we will not discuss these tasks here. They are all covered in Part III of the book, where these ideas are turned into actual working applications. The coverage may not be exactly as listed in this chapter, but the general ideas are preserved.

Part II

INPUT AND OUTPUT: THE BASIC TECHNIQUES TO BE MASTERED—LEARNING BY DOING

Although the idea is losing currency in today's academic world, we learn best by doing. Learning works best when the hands, the eyes, and the brain work together to reinforce one another. With this in mind, we will proceed with undertaking a number of progressively difficult experiments that lend themselves to implementation on the Propeller system. In this part of the book, we will develop the basic techniques necessary to use the Propeller to address any number of real-world situations. In Part III of the book, we will use what we have learned here to create some real-world devices. In this part of the book, we are mastering the building blocks.

12

GENERAL DISCUSSION OF INPUT/OUTPUT

This is a general discussion of how we will proceed with learning the basics of using the Propeller so you will understand the how and why of the experiments and thus get the most out of them.

The experiments we will undertake are designed to give you the hands-on experience you need to get familiar with the Propeller chip. As such, I have tried to make them as short and as easy to understand as possible. Within the various experiments, I tried to incorporate as many of the basic Spin language techniques as I could, to make you more comfortable with the software environment. As was stated previously, the Assembly language programming facility provided for the Propeller has not been used in any of the programs except for the prewritten code for reading an optical encoder used in Chapter 28 as a part of the DC motor experiment. This is a book specifically for beginners, and PASM is really too difficult to get into at this stage. Besides, you really need to have a pretty good feel for binary mathematics and Boolean algebra before you can use PASM because a lot of it has to do with manipulating bits and bytes, and a foundation is required before it can be used. (I considered the possibility of writing a beginner's book on PASM, but decided it was not really a subject for beginners!)

The experiments start off with the most rudimentary of all experiments: reading an input line. Because we have to have a way to verify that we actually read the input, we will turn an LED on if and only if the input line has been grounded. This experiment contains the code segments needed for an absolutely minimal program. In some ways, this is about as short an input/output program as one can write for the Propeller. This is the prototypical Spin program, and all programs are basically variations of this program. We will build on it from here on out.

The purpose of the first experiment is not so much to turn the LED on and off as it is to get familiar with everything you need to do to create a fully functional program that is downloaded to the Propeller chip and causes some sort of a response in the hardware that you can see. The easiest way to do this is with an LED. You will also notice that this object (program) incorporates simple methods (subroutines) to introduce you to how methods are used within the Spin language. It is not a parallel-programming example.

From there, we make a fairly big leap to using a liquid crystal display (LCD) with the Propeller and developing all the methods that we will need to use it in the experiments that follow. We have to make this jump because we need an LCD (or something else) to show us what is going on within our experiments in a visual format. All the methods we need to give us comprehensive access to the LCD are developed, and the techniques for storing them at a suitable location on disk and then calling them from within another object are covered. A comprehensive discussion of what one needs to do to control an LCD is provided in Chapter 14. The discussion there gives you the information you need to control all aspects of the operation of a typical 16×2 LCD from the absolute beginning.

Although writing to a computer display is mentioned in this tutorial, I do not cover the creation of the software needed to communicate with a display because the subject is more advanced than is suitable in a book for beginners. You should not let this issue keep you from using larger displays because there are a number of objects (in the object exchange) that allow you to use a display with minimal effort. These programs are in the public domain, and I encourage you to learn how to use them based on the use demonstrated for the LCD. The way you access a larger display is similar to how you access the LCD, and using an LCD is covered in detail.

Once we have an operational LCD, we can start to develop the techniques we need to bring information into the Propeller and to get output information from the Propeller. The LCD programs developed allow us to see what is going on within the system with minimal new programming. As we proceed, you will see that we really do need to be able to look into an operating program to see what is going on, and that an LCD can be quite adequate for doing that. Besides, the LCD is the most inexpensive way to get a self-contained operation going with the kit we are using.

The inputs and outputs we have the greatest interest in are the types of signals that computers use to interact with the world. A summary of them follows:

- Pulsing
- Using the LCD to see what is going on
- Binary I/O interaction
- Reading and creating pulse widths
- Reading and creating frequencies
- Reading and creating pulse sequences
- Read a varying DC voltage signal (generated by a potentiometer)

Generally, almost everything we read into a computer and send out of a computer comes in and goes out in one of these formats. Each of these is covered in a separate chapter to compartmentalize things and keep confusion to a minimum.

We need a source that provides us with signals that we can respond to. Because we are operating in a parallel-programming environment, the generation of any signals we may need can be assigned to one of the Propeller's cogs. We do not need a separate programmable signal generator to provide our signal needs. Because input/output is what it is all about, learning how to generate the signals we need is an important skill for us to master.

Mastering the preceding input and output techniques gives us a basic understanding of the processes used to get information in and out of a microprocessor. In that we have eight cogs to work with, we can assign one cog to read the information and another to put it out in the same shape and form as the basis for our experiments. If both the input and output waveforms look to be (somewhat!) identical in some respect, we will have successfully read and generated the waveform under consideration. You can use one trace of your oscilloscope to look at the incoming waveforms and the other to look at what you are sending out. There will be a delay between the two waveforms, and the shapes and frequencies will not be identical. However, they need to be pretty much similar if you want to claim success. There will also be some timing discrepancies, but we won't let that distract us. Let's see how well we do. Once we get things working, you can work on getting them perfect.

Once we are comfortable with the inputs and outputs covered in this part of the book, we will move to the third part of the book, where we will use what we have learned in Part II to create real-world devices. Part III of the book is dedicated to the construction of a number of devices that use the information we mastered in the first two parts of the book. All the projects are straightforward and are designed to be similar to the projects you might expect to undertake if you are interested in the real-world use of the Propeller chip.

13

BINARY PULSING

Before we can do any useful work with a microprocessor, we need a way to get information into and out of the processor. If nothing goes in and nothing comes out of the black box, its use is somewhat limited! The output has to be such that we can either read it or connect it to some other device that can respond to it. We or the device then reacts to this signal in a way that creates useful information or work. That, in its simplest form, is the application of computers to solve the problems we are interested in.

The simplest output any programmable device can provide is a signal that goes on and off. The rate at which the line goes high and low and the relative timing of the high and low signals can provide useful information in any number of ways. Most of the serial communications that computers undertake between one another is based on the manipulation of such signals, as defined by the ASCII codes. All the communication within the computer itself is undertaken with on/off signals. We run motors with PWM (pulse width modulated) signals that vary the duty cycle of the signal between zero and one, and thus the speed of the motor. More properly, the power to the motor is said to be a function of the PWM signal. Therefore, learning how to manage these on/off signals is pretty much fundamental.

For most purposes, it is not possible for human beings to use the information that the signal provides without some kind of secondary manipulation or amplification of the signal. The most common interface is the computer monitor. Learning the techniques for using the signals that computers provide is an important part of learning how to use computers. In this, the second part of the book, we learn the basics of how to read and generate the various signals that computers create and need to do useful work. Part III of this book is devoted to running experiments and making devices that use the techniques we developed in Part II.

For our first exercise, we will write a simple program to blink an LED on and off on an even "on/off" cycle. The purpose of the program is not to blink an LED but rather to go through all the procedures that need to be undertaken to write a complete program and run it. By doing this, we become familiar with the operation of the entire system. Because we are all beginners as far as the Propeller chip goes, we also need to get familiar with what a program written in Spin looks like and we need to get familiar with what the procedures for running a Spin program are.

We also need to start thinking about incorporating methods into the objects we write. In the Spin language, subroutines are called "methods." Although there is really no need to call methods in the LED-blinking program we are about to examine, it uses three very simple two-line methods to perform the functions that manipulate the LED signals. A listing of the program is provided in Program 13-1.

Program 13-1 Blinking an LED: Simple Method Calls

```
{{12 Sep 09    Harprit Sandhu
BlinkLED.spin
Propeller Tool Ver. 1.2.6
Chapter 13 Program 1

This program turns an LED ON and OFF, with a programmable set delay.
It demonstrates the use of methods in an absolutely minimal way.
The clock is running at 10 MHz.

Define the constants we will use.
There are no variables in this program.

}}
CON
  _CLKMODE=XTAL1 + PLL2X           'The system clock spec
  _XINFREQ = 5_000_000             'the crystal frequency

  inv_high      =0                 'define the inverted High state
  inv_low       =1                 'define the inverted Low state
  waitPeriod    =5_000_000         'about 1/2 sec switch cycle
  output_pin    =27

'High is defined as 0 and low is defined as a 1 because we are using an
'inverting buffer on the Propeller output.

PUB Go
  dira [output_pin]~~              'sets pin to an output line with ~~
  outa [output_pin]~~              'makes the pin high
  repeat                           'repeat forever, no number after repeat
     turnOff_LED                   'method call
     wait                          'method call
     turnOn_LED                    'method call
     wait                          'method call
```

(continued)

Program 13-1 Blinking an LED: Simple Method Calls (*continued*)

```
PRI turnOn_LED                      'method to set the LED line high
    outa[output_pin] :=inv_high 'line that actually sets the LED high

PRI turnOff_LED                     'method to set the LED line low
    outa[output_pin] :=inv_low  'line that actually sets the LED low

PRI wait                            'delay method
    waitCnt(waitPeriod + cnt)   'delay is specified by the waitPeriod
```

Here is what is going on in Program 13-1:

- Program 13-1 declares four constants and then starts a public procedure that calls the three private methods that control the LED and the timing delay. All are named to be easy to remember.
- The CON block states the constants that will be used in the program. In this particular case, we have four constants that define the operating parameters for the program.
- PUB Go, the main procedure, defines the pin we are going to use as an output and then repeats the main loop of the program. The three PRI (private) methods that are called provide the On/Off and delay functions used by the program.

Other programs you write may be much more complicated, but they will follow this basic layout. We will creep up on more complex programs so that you will have no problem following what is going on.

The wiring needed between the Propeller chip and the LED is shown in Figure 13-1. I have omitted all the power wiring so that we can concentrate on the wiring we have to create. The power wiring follows the recommendation of Parallax for their Education Kit, which will be used as the basic layout throughout this book. All the experiments will fit on the prescribed Education Kit assemblage of breadboards. The layout of the basic electronics that power the board is shown in Figure 13-2.

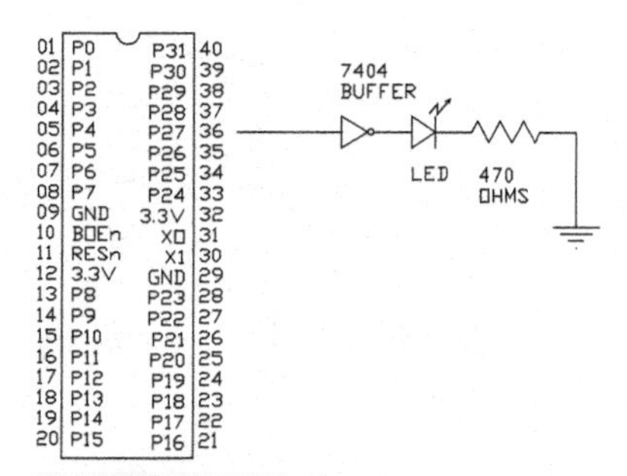

Figure 13-1 Wiring layout for blinking an LED

Once you have the wiring connected up and the program running, you will have learned the basic procedure for running all programs.

As a general rule, we will reserve pins P26 and P27 for general output from the Propeller chip. These lines need to go through a 7404 hex buffer to prevent overloading the very limited current capacity of the Propeller pins. Figure 13-3 shows the placement of the 7404 hex buffer on the breadboards to accomplish this.

For the record, Figure 13-2 provides the schematic for wiring up the basic Propeller layout. This is as suggested for the Propeller Education Kit by Parallax.

Note *This layout will be used for all the experiments in this book.*

The Memory module is not shown in this layout, and we will not need it for any of our experiments. However, you can install it if you like for your own experiments. All experiments use this basic layout. Parallax recommends that you connect power and ground to both sides of the Propeller, as is shown in Figure 13-2.

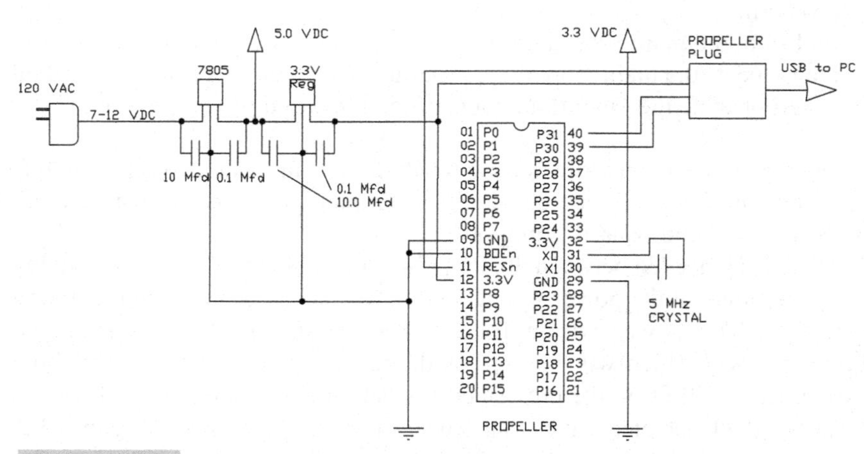

Figure 13-2 **Basic power-up layout and USB connection for a Propeller chip**

Figure 13-3 Hex buffer placement and wiring layout

14

SETTING UP A 16-CHARACTER-BY-2-LINE LIQUID CRYSTAL DISPLAY

Although this is early in the game, we absolutely have to have some way to look at the results of what we are doing in the Propeller as we develop our programs. We will go over the operation of a 16-character-by-2-line LCD in absolute detail later in Chapter 21. For now, we just need to design the wiring layout and develop the code that will allow us to start using the LCD with our experiments (with minimal explanations). The minimal discussion provided in this chapter is designed to keep you from getting completely lost as you install your LCD onto the breadboard.

Almost all the 16-character-by-2-line displays on the market are controlled by the Hitachi HD44780 or a compatible LCD controller. Do not purchase an LCD that does not meet this specification because all the work described in the book is based on LCDs as controlled by this Hitachi controller. The possibility that we could buy a bare LCD and write all the code to control it is beyond the scope of this book.

Looking through the data sheet that comes with the display, you will find a pinout table very similar to the one shown in Table 14-1. We will use this table to wire the LCD to the Propeller chip. We will be using the Parallax Education Kit wired exactly as suggested by Parallax for their educational materials. See Figure 14-1 for a pictorial representation of the connections.

Note *Many of the units do not support lines 15 and 16, and we will not be using them in any of our experiments either.*

Position the LCD on a bottom line of the board, as shown, with pin 14 in the rightmost holes at the bottom of the perforated board assembly. The wiring scheme we will be using is listed in Table 14-2.

Note *This scheme is suitable for 4-bit or 8-bit communications.*

TABLE 14-1 PINOUT CONNECTIONS FOR A TYPICAL 16×2 LCD

PIN NO.	SYMBOL	DESCRIPTION	
1	VSS	Logic ground.	
2	VDD	Logic power 5 volts.	
3	VO	Contrast of the display. Can usually be grounded.	
4	RS	Register select.	These are
5	R/W	Read/Write.	the three control
6	E	Enable.	lines.
7	DB0		
8	DB1		
9	DB2	This is	
10	DB3	one port or	
11	DB4	eight lines	Half the port
12	DB5	of data.	can also
13	DB6		be used.
14	DB7		See the data sheet.
15	BL	Backlight power.	These two lines
16	BL	Backlight ground.	can be ignored.

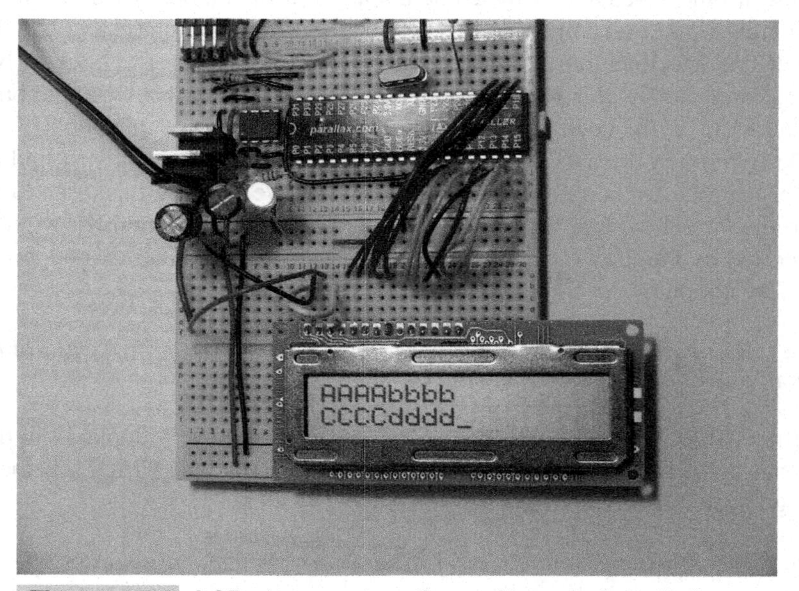

Figure 14-1 LCD placement on the perf board—installed at extreme right, as low as possible. (Eight-bit path shown.)

TABLE 14-2 WIRING CONNECTIONS BETWEEN THE PROPELLER AND THE LCD

PIN NO.	SYMBOL	PROPELLER PIN
1	VSS	Ground
2	VDD	5 volts
3	VO	Ground
4	RS	P18
5	R/W	P17
6	E	P16
7	DB0	P15
8	DB1	P14
9	DB2	P13
10	DB3	P12
11	DB4	P11
12	DB5	P10
13	DB6	P9
14	DB7	P8
~~15~~	~~BL~~	
~~16~~	~~BL~~	

Data can be transferred to the LCD either 8 bits at a time or 4 bits at a time. Initially, we will wire up all eight lines of the data bus, but once we get the LCD working, we will change over to 4-bit mode so that we can free up four lines for other uses.

The Spin code we will be using to access the LCD is listed in Program 14-1. This code will display four *A*'s and four *a*'s on line 1 and four *B*'s and four *b*'s on line 2 when everything is wired up correctly and the program is downloaded and run. Wire up the LCD and run the program to confirm this.

The code for this program is listed here so you can see what we are doing, but the code will not be explained at this time so that we can move on to the experiments as soon as possible. The various routines will be placed in an "LCDRoutines4" program, and we will be able to call every method in LCDRoutines4 by its name from our programs. We will not have to type in any of this code again. All this is explained in Chapter 21, which is devoted to a very detailed explanation of how an LCD needs to be set up and the programs written to access the features we have in mind.

The listing provided here also demonstrates the ability of the Spin language to create electronic diagrams within the Spin language with the Parallax font. Only the diagram for this exercise was drawn with the Parallax font in this book. All other diagrams were drawn in AutoCAD because at this time the Parallax font does not permit the creation of logic gates and some other elements that need to be documented for our experiments.

I understand that this deficiency is to be remedied in a future version of the chip and may well be in place by the time you read these words.

Program 14-1 Implementation of LCD Use for Our Experiments (Listing of Methods Available within "LCDRoutines")

```
{{11 Sep 09      Harprit Sandhu
LCDminimal.spin
Propeller Tool Ver 1.2.6
Chapter 14  Program 1

PROGRAM TO BEGIN USING THE LCD

A minimal LCD implementation to allow us to use the LCD in our
experiments immediately.

This program is an absolutely minimal implementation to
make the LCD usable. You can work on improving it. We will improve it later.
```

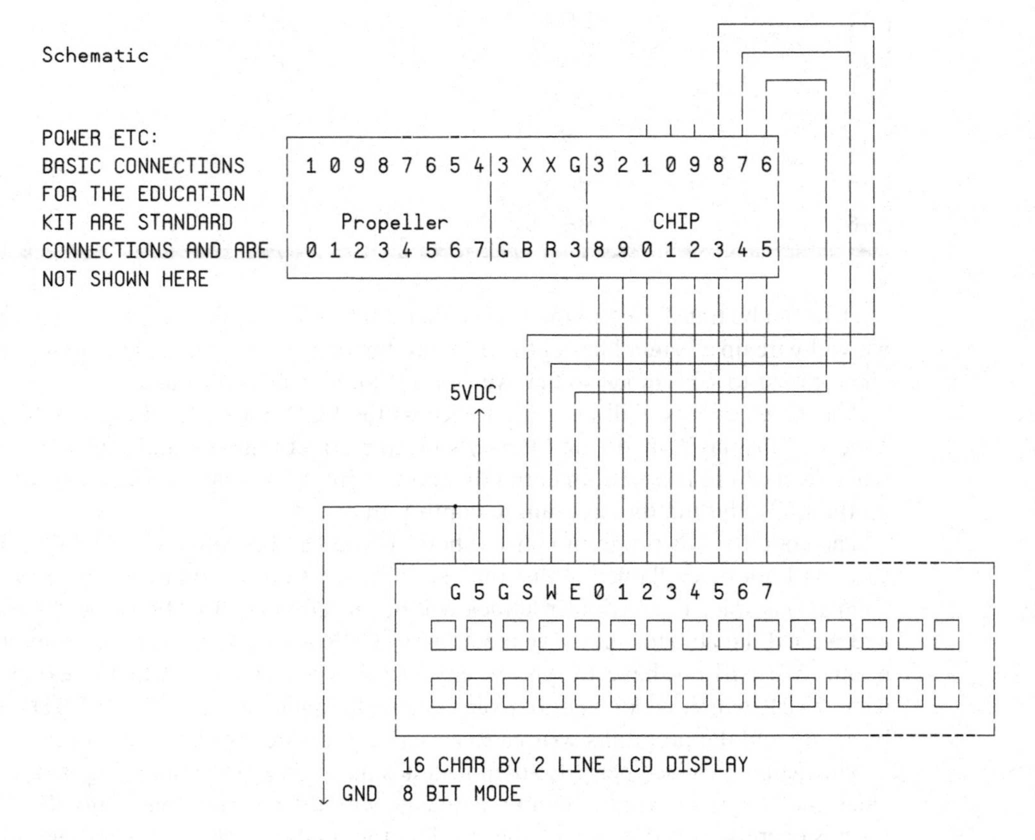


```
Revisions
```

(continued)

Program 14-1 Implementation of LCD Use for Our Experiments (Listing of Methods Available within "LCDRoutines") (*continued*)

```
Pin assignments are assigned as constants because the pins are fixed.
These numbers reflect the actual wiring on the board between the Propeller
and the 16x2 LCD display. If you want the LCD on other lines, that would
have to be specified here.  We are going to use 8-bit mode to transfer
data for now.  All these numbers refer to lines on the Propeller.
}}

CON
  _CLKMODE=XTAL1 + PLL2X          'The system clock spec
  _XINFREQ = 5_000_000            'the crystal frequency

  RegSelect     = 16
  ReadWrite     = 17
  Enable        = 18
  DataBit0      = 8
  DataBit7      = 15
  waitPeriod    =500_000                   'set the wait period
  high          =1                         'define the High state
  low           =0                         'define the Low state
{{
Defining high and low states will allow us to invert these when we use
buffers to amplify the output from the prop chip.  We will then make
low=1 and high=0 thus inverting all the values throughout the program.
}}

PUB Go
  DIRA[DataBit7..DataBit0]:=%11111111  'the lines for the LCD are outputs
  DIRA[RegSelect] := High          'the lines for the LCD are outputs
  DIRA[ReadWrite] := High          'the lines for the LCD are outputs
  DIRA[Enable]    := High          'the lines for the LCD are outputs

  INITIALIZE_LCD                   'initialize the LCD
  waitcnt(1_000_000+cnt)           'wait for LCD to start up
  CLEAR                            'clear the LCD
  repeat 4                         'print 4 'A's
    SEND_CHAR ("A")
  repeat 4                         'print 4 'a's
    SEND_CHAR ("b")
  POSITION (1,2)                   'move to POSITION: line 2, space 1
  repeat 4                         'print 4 'B's
    SEND_CHAR ("C")
  repeat 4                         'print 4 'b's
    SEND_CHAR ("d")
  repeat                      'this is a parking loop to keep the system
                              'from shutting down. It just loops here
                              'see what cursor does if it is omitted
```

(continued)

Program 14-1 Implementation of LCD Use for Our Experiments (Listing of Methods Available within "LCDRoutines") (*continued*)

```
PRI INITIALIZE_LCD              'The addresses and data used here are
  waitcnt(500_000+cnt)          'specified in the Hitachi data sheet for
                                'display. YOU MUST CHECK THIS FOR YOURSELF
  OUTA[RegSelect] := Low        'these three lines are specified to write
  OUTA[ReadWrite] := Low        'the initial set up bits for the LCD
  OUTA[Enable]    := Low        'See Hitachi HD44780 data sheet
                                'display. YOU MUST CHECK THIS FOR YOURSELF.
  SEND_INSTRUCTION (%0011_0000)    'Send 1st
  waitcnt(49_200+cnt)              'wait
  SEND_INSTRUCTION (%0011_0000)    'Send 2nd
  waitcnt(1_200+cnt)               'wait
  SEND_INSTRUCTION (%0011_0000)    'Send 3rd
  waitcnt(12_000+cnt)              'wait
  SEND_INSTRUCTION (%0011_1000)    'Sets DL=8 bits, N=2 lines, F=5x7 font
  SEND_INSTRUCTION (%0000_1111)    'Display on, Cursor on, Blink on
  SEND_INSTRUCTION (%0000_0001)    'clear LCD
  SEND_INSTRUCTION (%0000_0110)    'Move Cursor, Do not shift display

PUB CLEAR                           'Clear the LCD display and go home
  SEND_INSTRUCTION (%0000_0001)

PUB POSITION (LINE_NUMBER, HOR_POSITION) | CHAR_LOCATION  'Pos crsr
  'HOR_POSITION : Horizontal Position : 1 to 16
  'LINE_NUMBER : Line Number : 1 or 2
  CHAR_LOCATION := (HOR_POSITION-1) * 64   'figure location
  CHAR_LOCATION += (LINE_NUMBER-1) + 128   'figure location
  SEND_INSTRUCTION (CHAR_LOCATION)   'send the instr to position cursor

PUB SEND_CHAR (DISPLAY_DATA)   'set up for writing to the display
  CHECK_BUSY                   'wait for busy bit to clear before sending
  OUTA[ReadWrite] := Low       'Set up to read busy bit
  OUTA[RegSelect] := High      'Set up to read busy bit
  OUTA[Enable]    := High      'Set up to toggle bit H>L
  OUTA[DataBit7..DataBit0] := DISPLAY_DATA    'Ready to SEND data in
  OUTA[Enable]    := Low       'Toggle the bit H>L

PUB CHECK_BUSY | BUSY_BIT       'routine to check busy bit
  OUTA[ReadWrite] := High       'Set to read the busy bit
  OUTA[RegSelect] := Low        'Set to read the busy bit
  DIRA[DataBit7..DataBit0] := %0000_0000  'Set the entire port to  input
  REPEAT                        'Keep doing it till clear
    OUTA[Enable]  := High       'set to 1 to get ready to toggle H>L bit
    BUSY_BIT := INA[DataBit7]   'the busy bit is bit 7 of the byte read
    OUTA[Enable]  := Low        'make the enable bit go low for H>L toggle
  WHILE (BUSY_BIT == 1)         'do it as long as the busy bit is 1
  DIRA[DataBit7..DataBit0] := %1111_1111  'set the port back to outputs
```

(*continued*)

Program 14-1 Implementation of LCD Use for Our Experiments (Listing of Methods Available within "LCDRoutines") (*continued*)

```
PUB SEND_INSTRUCTION (DISPLAY_DATA) 'set up for writing instructions
  CHECK_BUSY                        'wait for busy bit to clear before sending
  OUTA[ReadWrite] := Low            'Set up to read busy bit
  OUTA[RegSelect] := Low            'Set up to read busy bit
  OUTA[Enable]    := High           'Set up to toggle bit H>L
  OUTA[DataBit7..DataBit0] := DISPLAY_DATA    'Ready to READ data in
  OUTA[Enable]    := Low             'Toggle the bit H>L
```

Program 14-1 will be improved upon, added to, and then broken up into its various methods as our skills improve. Once we have worked up a comprehensive set of methods, the methods will be placed in an object called LCDRoutines4, from where we will be able to call whatever method we need in any of the programs we develop. You do not have to worry about the details of this at this time. It will all be explained later in Chapter 21. The final versions of the programs LCDRoutines4 and Utilities are also listed in Appendix A. LCDRoutines4 is the 4-bit version of LCDRoutines.

The Spin language is designed to allow the easy sharing of previously developed programs by third-party users. You are welcome to use any of the software developed in this book for use in any way you see fit as per the terms of the MIT license described in the preface of this book. You should also develop some expertise in using other objects in the object exchange maintained by Parallax so that you do not have to reinvent the wheel from time to time. Get familiar with the object exchange. It is a very useful resource for beginners.

15

BINARY INPUT AND OUTPUT: READING A SWITCH AND TURNING ON AN LED IF THE SWITCH IS CLOSED

In any real-world situation, you'll be reading inputs either as discrete single-line inputs or as resistances, frequencies, pulse widths, or similar signals. In this chapter, we will consider reading one external input as being either high or low. Because we need to know whether we have actually read the input successfully, we will turn an LED on if and only if the input is read as being low. We will define one pin as an input for the input signal and one pin as an output to turn on the LED. All the work in this program is being done in one cog, although in a parallel-processing environment we could assign the input to one cog and the output to another, as we will do in many of the following programs.

First, let's take a look at the program and then we will discuss what each part of the program is doing. For our purposes, the signal will go between a logical 0 and logical 1 at the Propeller input pin.

For the Propeller with its CMOS circuitry, anything over half the 3.3 DC volts power (or 1.65 volts) is a seen as a high. The input impedance of the lines in the input state is so high that they tend to float and will drift between 0 and 1 if they are not connected to anything (when defined as inputs). You can observe this if you run a finger close to a free input line with a wire sticking in the air like an antenna. The stray capacitance (usually a positive charge) of your body provides enough current to switch the line to a high condition as you approach it. For us, this means that all input lines have to be tied high or low, with a high resistance when we use them as inputs (or defined as outputs and then made high or low). If you are going to pull the line low in your circuit to read it, tie it high with a 1 meg resistor, and if you are going to pull the line high to read it, tie it to ground with a 1 meg resistor.

Keeping all this in mind, Figure 15-1 shows the circuit for reading a line that is connected to a switch that grounds it.

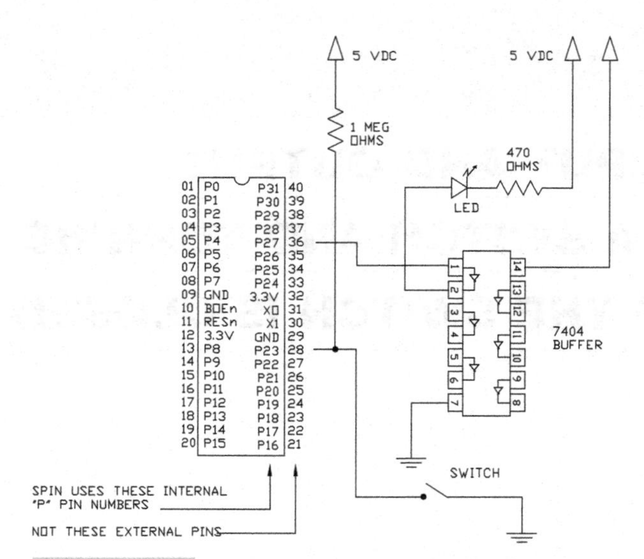

Figure 15-1 Switch controls LED on line P27 by pulling line P23 low.

Note *The power to drive the LED comes from the power to the 7404, not the Propeller. The 7404 inverts each signal it processes.*

In Figure 15-1, we are using a buffer in the 7404 to drive the LED. We are using a 1 meg resistor to pull line P23 up to 5 volts. The current to the LED is limited by the 470 ohm resistor. The switch grounds line P23. This is pin 28 on the chip. The Spin language uses the internal numbers, never the external chip pin numbers.

The code that runs the circuitry in Figure 15-1 is listed in Program 15-1. After you have studied it, we can discuss it.

Program 15-1 Reading a Switch and Turning an LED on while the Switch Is Down

```
{{12 Sep 09      Harprit Sandhu
ButtonLED.spin
Propeller Tool 1.2.6
Chapter 15 Program 1

This program turns an LED ON if an input that has been pulled up is
grounded It demonstrates the use of subroutines in an absolutely minimal
way.
First define the constants we will use.

}}
CON
  _CLKMODE=XTAL1+ PLL2X         'The system clock spec
  _XINFREQ      = 5_000_000
```

(continued)

Program 15-1 Reading a Switch and Turning an LED on while the Switch Is Down (*continued*)

```
  input_pin     =23             'select the pin to be used for input
  output_pin    =27             'select the pin to be used for the LED
  inv_high      =0              'define the inverted High state
  inv_low       =1              'define the inverted Low state

'High is defined as 0 and low is defined as a 1 because we are using an
'inverting buffer on the output

PUB Go
  dira [output_pin]~~    'sets pin to an output line with ~~ notation
  outa [output_pin]~     'makes the pin a low output
  dira [input_pin]~      'sets pin to an input line with the ~ notation
  repeat      'repeat forever because there is no number after repeat
   if ina[input_pin]==0     'check pin for high or low
     turnOn_LED          'subroutine call
   else
      turnOff_LED        'subroutine call

PRI turnOn_LED                        'subroutine to set the LED line high
    outa[output_pin] :=inv_high 'line that actually sets the LED high

PRI turnOff_LED                       'subroutine to set the LED line low
    outa[output_pin] :=inv_low  'line that actually sets the LED low
```

Discussion

This is essentially how we read a switch into the Propeller chip when we need to. If we were looking at extremely fast phenomena, we would have to take measures to de-bounce the switch in either hardware or software to make sure we did not misinterpret its operation. We will not worry about that at this time, but you should be aware of the fact that switch bounce is a problem with mechanical switches. Mechanical switches are hundreds, even thousands of times slower than even slow microcontrollers.

We defined all the constants at the top of the program. Defining constants allows you to make changes to the I/O line identifications and such with ease. When we do it in this way, any changes we make will be reflected throughout the program automatically.

We define a high as 0 and a low as 1 in this program because we are using one of the inverting hex buffers in the 7404. These buffers turn a 0 into a 1 and a 1 into a 0, so we take care of this inversion in the software definitions. Note that this does not affect the input at the input pin because we are reading that directly (there is no intermediate buffer). As a rule, we can connect to high impedance inputs directly, unless we are dealing with high voltage, in which case special safety precautions must be undertaken.

If you remove the pull-up register on the input line, you will notice that even putting an oscilloscope on the line will pull it down. As discussed previously, all input lines must be tied high or low if you want to prevent unexpected behavior.

The Repeat Command

When no number appears after the repeat command, the indented lines under the repeat command are repeated endlessly. If there is a number, it defines how many times the lines are to be repeated.

Note that we are using methods to perform simple tasks such as turning the LED on and off. Although not strictly necessary in this case, we are using methods to do even simple tasks now because in the long run this is the best way of doing it. We say it is "good practice." It allows things that we do often to be made easy, and the methods themselves are easy to edit. The changes you make to a method will be reflected throughout all the objects you develop, automatically, if you use these programming techniques.

16

READING A POTENTIOMETER: CREATING AN INPUT WE CAN VARY IN REAL TIME

Along with binary on/off switches, which provide a two-state input, we need to be able to enter information that we can vary with a rotating knob, so that we can see the effect of varying the inputs on our experiments. This is usually done with a potentiometer placed across a voltage. This being the case, we need to learn how to interact with such a voltage next. In this chapter we will learn how to read a potentiometer first into one byte (and later on with a resolution of 12 bits). One byte gives us a value between 0 and 255 for the full range of the potentiometer rotation. We can use this to control a variable with a resolution of approximately 0.39% (one part in 256). It is also possible to read a potentiometer into more than one byte; 12-bit analog-to-digital (A-to-D) converters provide an easy way to get much higher resolutions, and we will consider their use at the end of the chapter. Twelve bits provide a resolution of one part in 4,096, or 0.0244%.

Note *We will set up most of our experiments with the ability to use two variable inputs, meaning that we will need the ability to read two potentiometers into our experiments. This will allow us to vary two variables in real time when we need to without having to change the experimental setup. We will not discuss a two-potentiometer setup in this chapter, but our experimental setups will have the ability to use two variable inputs. Reading the second potentiometer is similar to reading the first one.*

One way to measure the resistance of a potentiometer without an A-to-D converter is to charge a capacitor to a known voltage and then discharge it through a resistor until it gets down to a known, specific voltage. The time it takes for the capacitor to discharge will be a function of the resistance of the potentiometer. The relationship is not strictly linear, but it is good enough for our immediate purposes.

When a Propeller line is high, it is at 3.3 volts. It switches from high to low at half this voltage, or 1.65 volts. These two facts are the basis for the resistance determination. Our task is to measure the time it takes for the capacitor to go from 3.3 volts to 1.65 volts. We do this by connecting the capacitor to a line of the Propeller. In our program we are using line P19. By making this line an output (high) and waiting a few milliseconds, we charge the capacitor attached to it to 3.3 volts. We then turn the line into an input and turn on a timer immediately. We monitor the state of the input pin. As soon as it goes low (at 1.65 volts), we read the timer again. Because we know the speed of the oscillator, as well as the time it took for the voltage to drop to 1.65 volts, we can get a reading that is related to the value of the resistance.

We read the counter twice, immediately after turning the line into an input and immediately after the line switches from high to low. The difference is related to the resistance. The lower the resistance, the lower the time difference. However, even at the lowest resistance it takes time for the various program instructions to execute. This minimum value has to be subtracted from the difference and represents the adjusted shortest time. The highest count read with the potentiometer at one extreme is divided by a number that produces a value of 255. The timer reading at the other extreme potentiometer position represents the various delays the execution of the program instructions create. This value has to be subtracted from the delay count to get the 0 value for the potentiometer.

The value you pick for the capacitor should not be too small or too large. If it's too small, you will not get a high enough count to allow a good reading on the potentiometer. It will be too coarse. If too large a value is picked, the conversion will take too long. I used a 10K potentiometer and a 10 mfd (micro farad) capacitor with the circuitry shown in Figure 16-1 for Program 16-1.

Analog Inputs

READING A POTENTIOMETER

Assume that we will use a potentiometer with a maximum value of 10K ohms for this experiment. We will use the potentiometer as a voltage divider and read the voltage at the wiper. The voltage represents the analog value we are interested in.

For our first exercise, let's take a look at what we would need to be able to take the potentiometer reading with the Propeller chip (see Figure 16-1). In order to do this, we have to set up a simple circuit to charge and then discharge a capacitor. We determine the reading of the potentiometer from the information received from the circuit. Looking at the pin with an oscilloscope as we take the readings is very instructive. The next section explains this further.

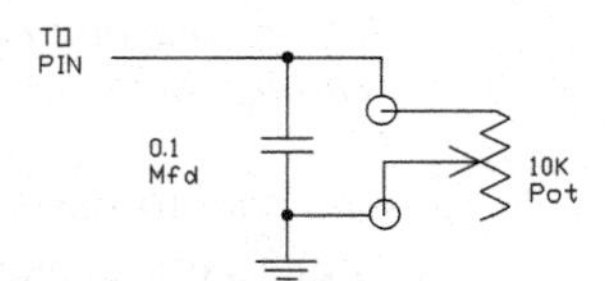

Figure 16-1 **Circuitry for reading a potentiometer with a Propeller chip**

THE DETAILS

If we charge a capacitor to an arbitrary voltage and then slowly discharge it through a resistor across the capacitor, the time that it takes for the voltage to come down to any specific voltage from the full-charge voltage will be a function of the resistance through which the charge on the capacitor is being drained and the voltage driving the reaction. The apparatus is attached to a port of the Propeller that has been configured to be an output and then set high. In a few moments, this will charge the capacitor to the voltage that the microprocessor provides as the high signal on an output line (3.3 volts DC for the Propeller).

Once the capacitor is fully charged, the port is turned into an input, a timer is started, and the system is programmed to monitor the state of the port as the capacitor discharges. As expected, initially the input port will be read as being high. As the capacitor is discharged by the potentiometer, a point will be reached when the port will be read as being low. How soon this happens depends on the resistance that the potentiometer is set to. If we measure the high-to-low time interval with the timer we started, the interval will be a function of the resistance of the potentiometer. This phenomena is used to determine the value of the potentiometer setting.

We can assume that if the potentiometer is set to 0 ohms, the capacitor will discharge immediately and the timer will indicate a 0 time interval. If, on the other hand, we set the potentiometer to its maximum resistance, which in our case is 10K ohms, the time of discharge will represent the maximum resistance of the potentiometer. All other values are represented by times between 0 and the maximum time. The relationship is not perfectly linear, but it's linear enough for all practical purposes. For our immediate needs we can consider it to be linear. This is expressed graphically in Figure 16-2.

For the sake of technical correctness, the rate at which a capacitor charges and discharges is more accurately represented in Figure 16-3.

We can assume a linear relationship because we are seeing a very tiny portion of this curve in our experiment, and for all practical purposes this tiny portion can be looked at as a straight line.

We can convert the relationship we have observed to provide a value of between 0 and 255 if we are interested in an 8-bit representation of the resistance of the potentiometer. The value of the capacitor we choose to charge is not critical, but too small a capacitor

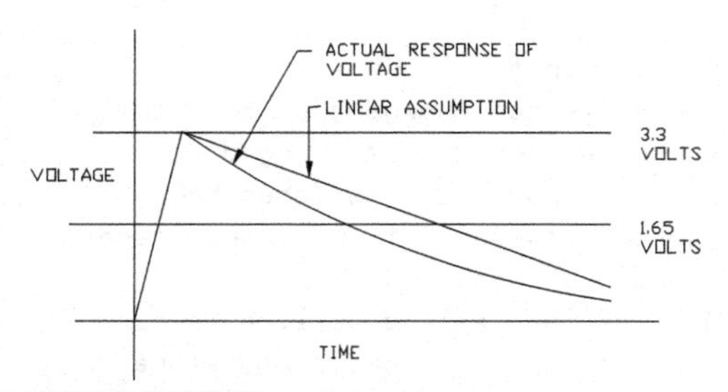

Figure 16-2 Graphic of resistance vs. time to discharge

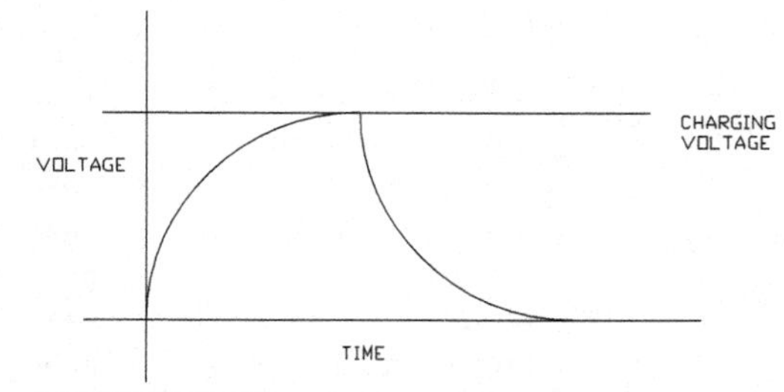

Figure 16-3 Theoretical charging and discharging of a capacitor

will give us insufficient time for the signals to settle down and too large a capacitor will take too long to discharge. In our particular case, a capacitor of 10 micro farads will give us an answer well within 0.01 seconds. This is fast enough for most purposes. Program 16-1 shows how the potentiometer was read, converted to an 8-bit value, and displayed on the LCD.

Program 16-1 Reading a Potentiometer

```
{{Aug 31 09   Harprit Sandhu
ReadPot.spin
Propeller Tool Version 1.2.6
Chapter 16 Program 1

READING A POTENTIOMETER

This routine reads a 10K pot with a 10 mfd cap.
This routine is what is used in the utilities to read the pot.
Pot is always read from the same line.

}}
CON
  _CLKMODE=XTAL1+ PLL2X           'The system clock spec
  _XINFREQ = 5_000_000            'Crystal spec
  PotLine   = 19                 'line the pot it on

OBJ
  LCD : "LCDRoutines4"        'We will be using these METHODS in this program

VAR                             'these are the variables we will use.
   long  startCnt               'count at start
   long  endCount               'count at end
   long  delay                  'time difference
   long  PotValue               'Value of the pot reading

PUB Go
  LCD.INITIALIZE_LCD           'set up the LCD
  repeat                       'loop
    dira[PotLine]~~            'set potline as output
    outa[PotLine]~~            'make it high so we can charge the capacitor
    waitcnt(4000+cnt)          'wait for the capacitor to get charged
    dira[PotLine]~             'make potline an input. line switches H>L
    startCnt:=cnt              'read the counter at start of cycle and store
    repeat                     'go into an endless loop
    while ina[PotLine]~~       'keep doing it as long as the potline is high
    EndCount := cnt            'read the counter at end of cycle and store
    delay := ((EndCount-StartCnt)-1184)    'calc time for line to go H>L
    if delay>630000                  'max permitted delay
      delay:=630000                  'clamp delay
```

(continued)

Program 16-1 Reading a Potentiometer (*continued*)

```
    PotValue:=(delay/2220)       'This reduces the value to 0-255 or 1 byte
    PotValue <#=255              'clamp range
    PotValue #>=1                'clamp range
    LCD.POSITION (1,1)             'Go to 1st line 1st space
    LCD.PRINT(STRING("PotPos =")) 'Potentiometer position
    LCD.PRINT_DEC(PotValue)        'print value
    LCD.SPACE(3)                   'erase over overflows
    LCD.POSITION (2,1)             'Go to 2nd line 1st space
    LCD.PRINT(STRING("Delay  =")) 'Print
    LCD.PRINT_DEC(delay)           'print value
    LCD.SPACE(3)                   'erase over overflows
```

Note *It may be necessary to adjust various constants used in the object to suit your particular setup.*

In Program 16-2, we are using a method in the Utilities object to read the potentiometer. This is the power of the Spin language. Once we have written a method, we can use it again and again, with just a couple of lines of code, in any object we write. The circuitry for Program 16-2 is shown in Figure 16-4.

The code that resides in the Utilities program at this time is shown in Program 16-2. The Utilities object will be added to from time to time as we get further along in the book.

Program 16-2 Code Segment to Read a Potentiometer from the Utilities Object

```
PUB GetPotVal
  dira[PotLine]~~             'set potline as output
  valutotal:=0                'clear total
    repeat  repval            'repeat
      dira[PotLine]~~         'set potline as output
      outa[PotLine]~~         'make it high so we can charge the capacitor
      waitcnt(4000+cnt)       'wait for the capacitor to get charged
      dira[PotLine]~          'make potline an input. line switches H>L
      startCnt:=cnt           'read the counter at start of cycle and store
      repeat                  'go into an endless loop
      while ina[PotLine]~~    'keep doing it as long as the potline is high
      EndCount := cnt         'read the counter at end of cycle and store
      delay := ((EndCount-StartCnt)-1184)    'calc time for line to go H>L
      if delay>610_000              'max permitted delay
        delay:=610_000              'clamp delay
      PotValue:=(delay/2300)        'Reduces value to 0-255 or 1 byte
      valutotal:=valutotal+potvalue     'figures total
      potvalue:=valutotal/repval    'figure average
      potvalue <#=255
      potvalue #>=0
  result:=PotValue                  'figure average
```

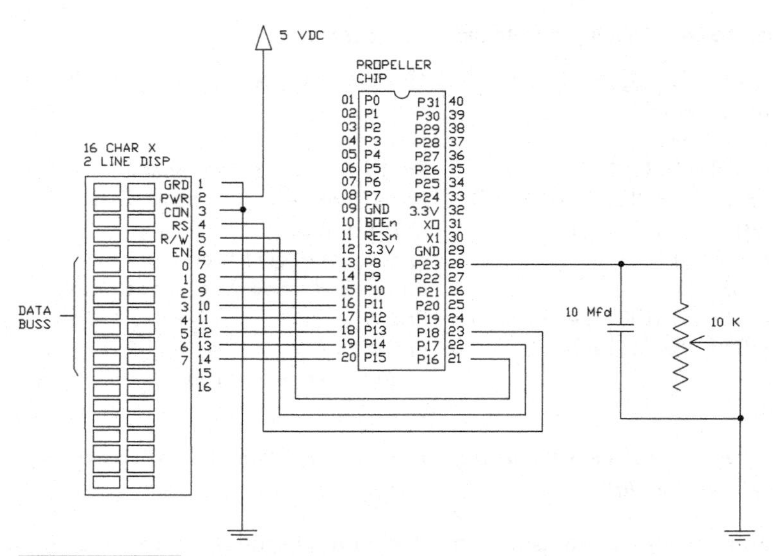

Figure 16-4 Complete circuitry for reading a potentiometer

We will not use the technique developed in these programs to read a potentiometer once this technique has been demonstrated. Instead, we will use an A-to-D converter to read the potentiometer(s). This A-to-D converter reads the potentiometers with a resolution of 12 bits to give us a reading between 0 and 4,095. This higher resolution is much more flexible for our purposes and will be used in all the following developments, as detailed next.

Advanced Techniques

USING THE MCP3202/MCP3208 FAMILY OF A-TO-D CONVERTERS

There are times when we need to get really serious about reading in a variable to a high resolution. If the variable can be expressed as a voltage between 0 and 5 volts, it can be read into the Propeller to a resolution of 12 bits with a chip identified as the MCP3202. (The 3208 is an eight-line version of the 3202.) This chip allows us to read a channel very rapidly, at 100,000 cycles per second (cps), and is available in versions that read two, four, and eight channels. It is a single-wire device. All these chips need a four-wire interface that meets the Serial Peripheral Interface (SPI) standard (go to the Internet to read more on the SPI standard).

The MCP3202 and MCP3208 are shown in Figure 16-5. Chips that read more than two channels need to have more pins to accommodate the additional channels.

Before you consider using the MCP3202 or the MCP3208 in any of your projects, download the data sheet to see if you can figure out what needs to be done to use the chip on your own. *This is a particularly good opportunity to get familiar with data sheets* because this is a simple chip that is relatively easy to understand and interface to. This ideal first-time opportunity should not be wasted by us beginners. I will go over the details, but I strongly recommend that you try figuring it all out on your own to gain learning experience and confidence.

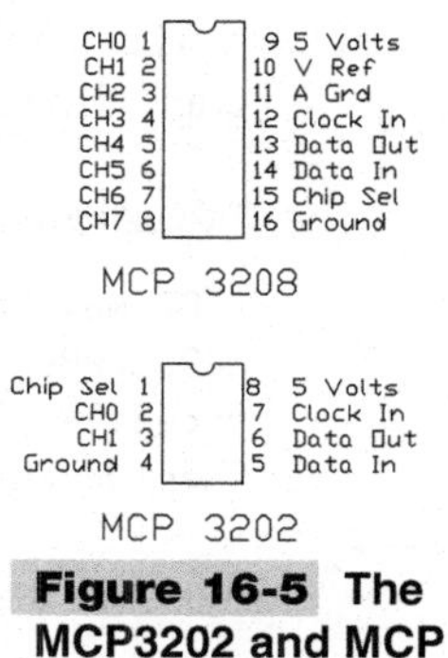

Figure 16-5 The MCP3202 and MCP 3208 pinouts

Because the chip follows the SPI standard for serial peripheral chips, we talk to it in a serial format, as specified in the data sheet. Here is what we have to do:

1. Get the wiring in place, as shown in Figure 16-7, later in this chapter.
2. Select the chip to make it active. This is done by pulling the Chip select line low from a high condition. If it is starting up in a low state, you have to make it high and then low to activate the startup sequence.
3. Tell the chip what mode we want to use it in. Four bits have to be either set or cleared. The four bits are toggled in one after the other and they specify the following conditions:
   a. To get going, we send the DataIn line a high "start bit."
   b. Next, we specify whether we are using single or differential mode.
   c. We need to specify which channel we want to use.
   d. We need to specify whether we want the LSB or the MSB read first.

Once this has been done, we are ready to read in the information to a variable that will store what we read.

After receiving these bits, the 3202 sends out a low (null) bit. We read it and discard it. The next 12 bits are the data we are interested in. Here's what happens:

1. Each bit is read in by making the clock go high to low.
2. Each time the clock goes low, the next bit of data becomes ready on the DataOut line.
3. We read the data into our variable.
4. We shift the bits in our variable left by one bit to make room for the next bit to be read in.
5. We then make the clock go high and then low, and the next bit becomes ready to read in.
6. We have to do all this 12 times to get all the 12 bits read in.

The detailed code for reading the 3202 is shown in Program 16-3. This particular program also serves as an example of full documentation and line-by-line commenting.

Program 16-3 Reading the MCP3202 Analog-to-Digital Chip to a 12-Bit Resolution

```
{{03 Nov 09   Harprit Sandhu
MCP3202Read1.Spin
Propeller Tool 1.2.6
Chapter 16 Program 3

All the code in this program is in Spin.

This program reads channel 0 of the MCP3202 and displays the results
on the LCD both as a decimal value and as a binary value so that you
can see the bits flip as you turn the potentiometer.

The 3202 chip is connected as follows:
1 Chip select                                       P21
2 Channel 0 for voltage input from Pot              Pot wiper
3 Channel 1 for voltage input from Pot, not used    Ground it
4 Ground Vss                                        Ground
5 Data into 3202 for setup                          P19
6 Data out from 3202 to be read into Propeller      P22
7 Clock to read in the data                         P20
8 Power 5 volts Vdd                                 5 volts

The Potentiometer is connected as follows:
Left    Ground
Center  To pin 2 of the 3202
Right   Power  5 volts
I used a 50K Pot

The connections to the LCD are as follows:
1   Ground
2   Power 5 volts
3   Ground
4   P16
5   P17
6   P18
7   Not connected, using 4 bit mode for data Xfer
8   Not connected, using 4 bit mode for data Xfer
9   Not connected, using 4 bit mode for data Xfer
10  Not connected, using 4 bit mode for data Xfer
11  Data  high nibble
12  Data  high nibble
13  Data  high nibble
14  Data  high nibble
```

(continued)

Program 16-3 Reading the MCP3202 Analog-to-Digital Chip to a 12-Bit Resolution (*continued*)

```
STANDARD EDUCATION KIT SET UP.  Used as base

Revisions:

Error Reporting:
Please report errors to harprit.sandhu@gmail.com

}}
OBJ
  LCD      : "LCDRoutines4" 'for the LCD methods

CON
  _CLKMODE=XTAL1+ PLL2X           'The system clock spec
  _XINFREQ = 5_000_000            'crystal spec
   BitsRead =12
   chipSel  = 19
   chipClk  = chipSel+1
   chipDout = chipSel+2
   chipDin  = chipSel+3

VAR
  long stack2[25]
  word PotReading
  word DataRead

PUB Go
  cognew(Cog_LCD, @stack2)
  DIRA[0..7]~
  DIRA[chipSel]~~     'osc once to set up 3202
  DIRA[chipDin]~~     'data set up to the chip
  DIRA[chipDout]~     'data from the chip to the Propeller
  DIRA[chipClk]~~     'oscillates to read in data from internals
  repeat
    DataRead:=0               'Clear out old data
    outa[chipSel]~~           'Chip select has to be high to start off
    outa[chipSel]~            'Go low to start process

    outa[chipClk]~            'Clock MUST be low to load data
    outa[chipDin]~~           'must start with Din high to set up 3202
    outa[chipClk]~~           'Clock high to read data in

    outa[chipClk]~            'Low to load
    outa[chipDin]~~           'High single mode
    outa[chipClk]~~           'High to read
```

(continued)

Program 16-3 Reading the MCP3202 Analog-to-Digital Chip to a 12-Bit Resolution (*continued*)

```
    outa[chipClk]~              'Low to load
    outa[chipDin]~              'Odd = low channel 0
    outa[chipClk]~~             'High to read

    outa[chipClk]~              'Low to load
    outa[chipDin]~~             'MSBF high = MSB first
    outa[chipClk]~~             'High to read

    outa[chipDin]~              'making line low for rest of cycle
    outa[chipClk]~              'Low to load Read/discard the null bit
    outa[chipClk]~~             'High to read

    repeat BitsRead              'Reads the data into DataRead in 12 steps
      DataRead <<= 1             'Move data by shifting left 1 bit.
                                 'Ready for next bit
      outa[chipClk]~             'Low to load
      DataRead:=DataRead+ina[chipDout]  'Xfer the data from pin chipDout
      outa[chipClk]~~            'High to read
    outa[chipSel]~~              'Put chip to sleep, for low power
    PotReading:=DataRead         'Finished data read for display

PRI cog_LCD                          'manage the LCD
  LCD.INITIALIZE_LCD                 'initialize the LCD
  repeat
    LCD.POSITION (1,1)               'Go to 1st line 1st space
    LCD.PRINT(STRING("Pot=" ))       'Print Label
    LCD.PRINT_DEC(PotReading)        'print decimal value
    LCD.SPACE(4)                     'erase over old data
    LCD.POSITION (2,1)               'Go to 2nd line 1st space
    LCD.PRINT_BIN(PotReading,BitsRead)   'Print it as bits.
```

The ability to read with a resolution of 12 bits gives you a lot more flexibility than reading to 8 bits. Essentially, it means that a potentiometer that rotates a little over 270 degrees can be converted into 4,096 discrete values like an expensive high-count encoder. It gives us the ability to read the potentiometer with a resolution of ~270/4,096=0.0659 degrees. That's a high-resolution encoder and would cost well over $50. We can now make one for about $1.75 with a potentiometer. (The linearity of carbon film potentiometers is not very good so they cannot be used where linearity is important. If you need better linearity for your application, use a more expensive wire-wound potentiometer, even a multiturn one.)

Because the MCP3208 can read eight channels, some complicated robotic possibilities come to mind using analog instead of digital sensing for joints and such. A few potentiometers placed here and there could collect a lot of accurate positional information in a hurry.

Note *The wiring to the LCD is not shown here. It's the same as always.*

Because we will be using two potentiometers to read data into our experiments in real time, in many of our experiments it will be worth our while to make up a little module that can be plugged into the education kit whenever needed to allow us to read two potentiometers. The module I made up is shown in Figure 16-6. The wiring for the module is shown in Figure 16-7.

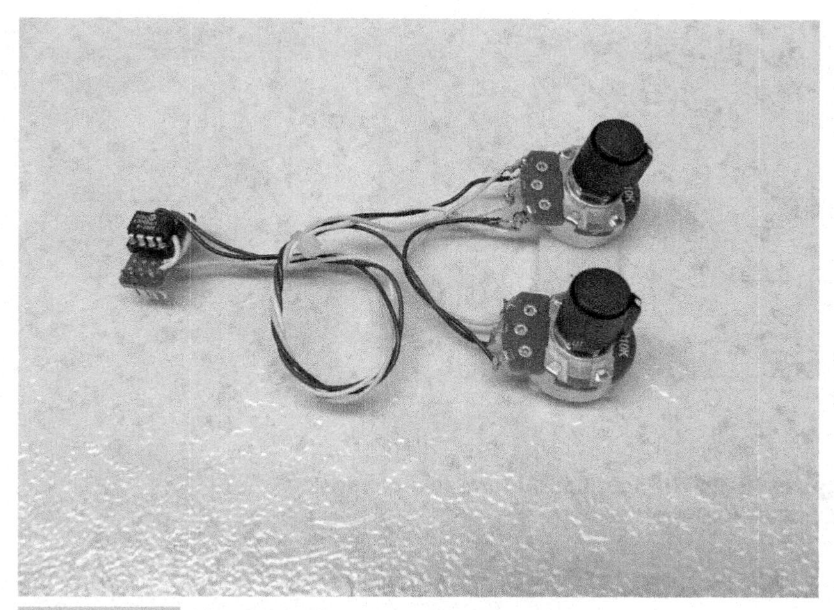

Figure 16-6 **The potentiometer reading module**

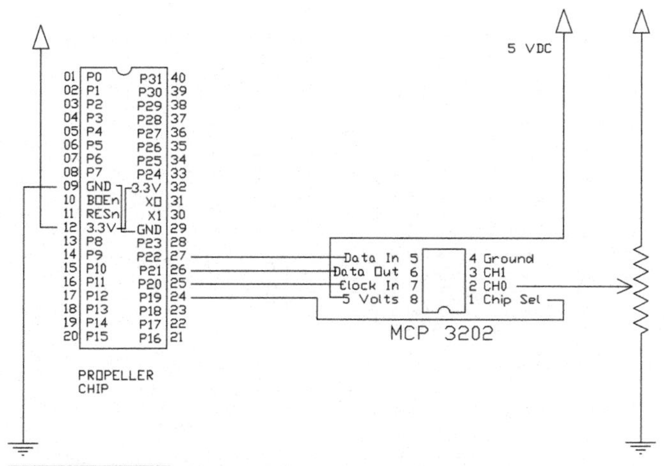

Figure 16-7 **Wiring diagram for the code shown in Program 16-3**

We will add the code to read the two potentiometers to the Utilities object as two independent methods for the two potentiometers. This will allow us to read each potentiometer independently to whatever resolution we desire, with 8 bits and 12 bits being the most frequently used. Later on, if we want to modify one of the methods to read one of the potentiometers in some different way, we will be able to do so. The wiring diagrams for using the MCP3202 and the MCP3208 are given in Figures 16-8 and 16-9.

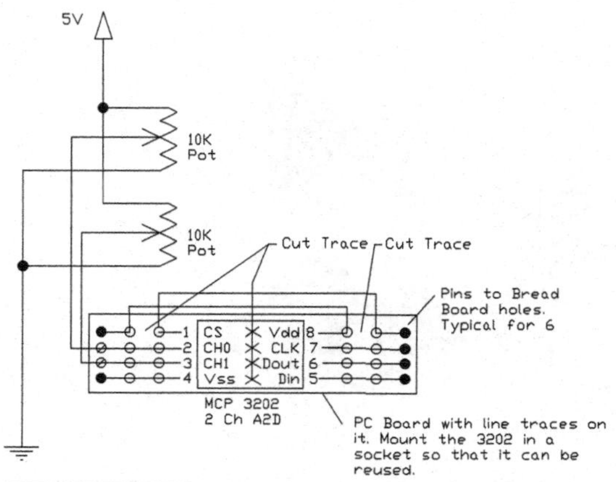

Figure 16-8 Wiring diagram for a two-channel MCP3202 A2D module

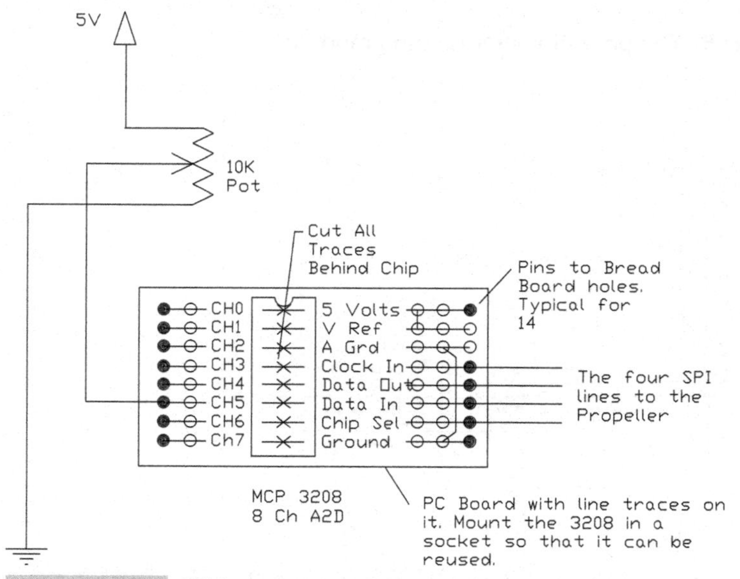

Figure 16-9 Wiring diagram for an eight-channel 3208 A2D module

Rule *From here on out, whenever we need to read a potentiometer (or anything else), we will use the MCP3202 A-to-D module as our peripheral chip to do so. We will always read the device to a resolution of 12 bits so that we will always get a value between 0 and 4,095.*

Note *A photograph that supports this wiring is shown in Figure 16-6.*

Note *Two channels are shown, each represented by one potentiometer.*

When you need to read two lines, you can use the wiring setup shown for an MCP3202 in Figure 16-10, later in this chapter. Only one potentiometer is shown there, but two can be read. The code for the eight-line 3208 chip is not given but can be easily developed from the code for the 3202 (see the data sheet). You have to use three clock/read cycles in the code so that eight lines can be specified.

Note *When using an A2D chip, you cannot connect the input channels to the input lines of the Propeller. They must be kept completely free of any extraneous load whatsoever if they are to read accurately. The output from the A2D as read by the Propeller depends on which line is specified as being interrogated. Thus, the four SPI lines to the Propeller can read all eight lines on an MCP3208.*

USING THE POTENTIOMETER READING IN THE LED BLINK PROGRAM

This section introduces the use of the waitCnt command (as modified by the potentiometer reading) to control the rate at which an LED blinks when we ground an input line. We will use the same setup as was used for Program 16-1 and will modify the software to do this. The wait method uses the waitCnt command to determine how the LED cycles on and off when the switch is pressed. The purpose of this program is to demonstrate the use of the waitCnt command.

The waitCnt command pauses the system until the system clock matches the argument of the command. The command is written as

```
waitCnt(waitPeriod + cnt)
```

where waitPeriod is the delay desired in clock cycles and cnt is the current reading of the 32-bit system counter or clock.

The instruction must be used exactly as written here.

Note *WaitPeriod must be more than 381 cycles of the clock because of the time it takes to execute the command. Any number below 381 is likely to stall the system or give unpredictable results.*

The waitCnt command uses the system clock to delay the program by a number expressed in clock cycles. The system clock is a 32-bit counter that increments by 1 during every clock cycle. The number of times the clock overflows is *not* kept track

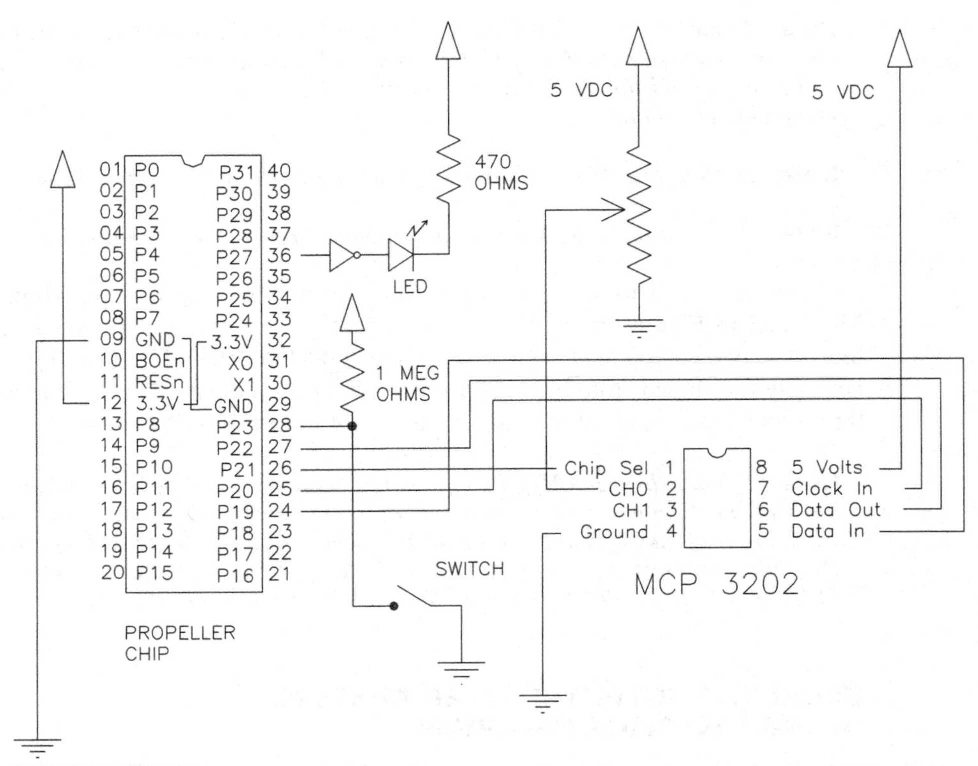

Figure 16-10 Wiring schematic for blinking an LED and reading a potentiometer

of by the system. The instruction only uses the difference in the counts to create the delays. If a delay that is more then 32 bits is specified, a problem will occur that has to be handled by the programmer. Delays expressed by numbers that exceed 32 bits are very unlikely to be needed within our programs.

The wiring schematic for this program is illustrated in Figure 16-10; the program for blinking the LED as controlled by the potentiometer delay is listed in Program 16-4.

Program 16-4 Blinking an LED at a Rate Controlled by a Potentiometer

```
{{12 Sep 09    Harprit Sandhu
BlinkLEDpot.spin
Propeller Tool Ver. 1.2.6
Chapter 16 Program 4

This program turns an LED ON and OFF, with a pot set delay.

Define the constants we will use.
}}
CON
```

(continued)

Program 16-4 Blinking an LED at a Rate Controlled by a Potentiometer (*continued*)

```
  _CLKMODE=XTAL1 + PLL2X           'The system clock spec
  _XINFREQ = 5_000_000             'the crystal frequency

  inv_high      =0                 'define the inverted High state
  inv_low       =1                 'define the inverted Low state
  output_pin    =27

'High is defined as 0 and low is defined as a 1 because we are using an
'inverting buffer on the Propeller output.

VAR
  long   Potvalue
  long   WaitPeriod
  byte   div                'dividing factor for clock freq

OBJ                         'These are the methods we will need
  LCD  : "LCDRoutines4"     'for the LCD methods
  UTIL : "Utilities"        'for general methods

PUB Go
  dira [output_pin]~~              'sets pin to an output line with ~~
  outa [output_pin]~~              'makes the pin high
  LCD.INITIALIZE_LCD               'initialize the LCD
  repeat                           'repeat forever, no number after repeat
     PotValue:=UTIL.read3202_0     'reads the potentiometer
     DIV:=1+PotValue/64            'adding 1 keeps value from going to zero
     turnOff_LED                   'method call
     wait                          'method call
     turnOn_LED                    'method call
     wait                          'method call
     LCD.POSITION (1,1)              'Go to 1st line 1st space
     LCD.PRINT(STRING("Pot=" ))      'dividing value
     LCD.PRINT_DEC(PotValue)         'print value
     LCD.SPACE(2)                    'erase over old data
     LCD.POSITION (2,1)              'Go to 1st line 1st space
     LCD.PRINT(STRING("Div=" ))      'dividing value
     LCD.PRINT_DEC(div)              'print value
     LCD.SPACE(2)                    'erase over old data

PRI turnOn_LED                      'method to set the LED line high
    outa[output_pin] :=inv_high     'line that actually sets the LED high

PRI turnOff_LED                     'method to set the LED line low
    outa[output_pin] :=inv_low      'line that actually sets the LED low

PRI wait                            'delay method
    waitCnt(clkfreq/div + cnt)      'delay is specified by the div
```

17

CREATING AND READING FREQUENCIES

For any number of reasons, we need to be able to create a signal at a given frequency and read the frequency of an incoming signal. This is one of the ways of sending information between all sorts of electronic devices. High-frequency signals are used as carriers for everything from radio, telephone, and TV transmissions to recording data on tape and on disk drives. Both the frequency and the amplitude of the signal can be varied. It is generally agreed that frequency modulation is more reliable and noise free than amplitude modification. For this reason, FM stations are clearer than the AM stations on your radio. However, FM does require considerably more bandwidth.

Our interest in this chapter concerns the ability to create and read specific frequencies as might be needed for the project we are working on. Once we know how to deal with the frequencies themselves, we can extend our skills related to using them as might be needed by the projects in Part III of the book.

In the experiment in this chapter, we will be working with audible frequencies so that we will not need any special instruments to detect the frequency. For the purpose of this experiment, just hearing a tone will suffice to indicate its presence. On the output side, we will create easy-to-recognize tones that emulate a musical progression. Because we will be creating rudimentary square waves, the sounds we create might leave something to be desired to a trained musician's ear, but they will suffice for our needs. We will be using the MCP3202 A2D converter to read a potentiometer, so you need to have this set up on your education board as was explained at the end of Chapter 16.

Creating Audible Frequencies

It is often necessary to create a specific frequency to meet a control or other experimental requirement. We can use the generation of musical tones with square waves as an exercise for learning how to generate frequencies. We'll be generating eight standard notes, starting with the A4 at 440 cycles per second and going up from there (it may not be exactly instrument-grade 440 cps but will be an acceptable tone for our exercise). The Propeller is set to run at 10 MHz because we are using a 5 MHz crystal and a multiplier of two. We will assume that the logic engine is running at exactly 10 MHz for the purpose of this discussion and the experiment (it will be very close).

Frequencies are generated by toggling a line up and down at a fixed rate. We will use equal-length high and low segments and vary the cycle time to get the note we want. With a processor running at 10 MHz, a 440 cps signal needs to toggle the line 440×2 times a second, or every 10,000,000/880 cycles of the system counter (10,000,000/880=11,364 cycles).

The frequencies for the eight notes we are interested in are as follows:

	Tone	Freq	Value Used
1	A4	440.00	440
2	B4	493.883	494
3	C5	523.251	523
4	D5	587.330	587
5	E5	659.255	659
6	F5	698.456	698
7	G5	783.991	784
8	A5	880.00	880

Notes *The exact frequencies are the 12th root of 2 apart.*
The 12th root of 2 is 1.05946309.
A4 is now standardized at 440 Hz. The standard has varied from about 420 to 450 Hz over the last three centuries.

The code segment shown in Program 17-1 will generate a 440 cps note (10,000,000/(440*2)=11,364).

Program 17-1 Code Segment for an A4 Note

```
dira[output]~~                       'output pin
repeat                               'loop
    !OUTA[output]                    'toggle output line
    waitcnt(11_364 +cnt)             'wait for the A4 freq
```

We need to generate this tone for about a quarter second and then go on to the next tone. We can repeat this code for as many tones as we need as shown in the preceding program segment. This code is incorporated into Program 17-2.

Program 17-2 Code for Generating Audible Tones

```
{{12 Sep 09    Harprit Sandhu
Play8Notes.spin
Propeller Tool Ver. 1.2.6
Chapter 17 Program 2

This program plays 8 notes, bare bones.

Define the constants we will use.
}}

CON
  _CLKMODE=XTAL1 + PLL2X              'The system clock spec
  _XINFREQ  = 5_000_000                'the crystal frequency
  repfactor =500                       'number of times repeated
  delayfact =1_000_000                 'delay between notes
  output    = 25

PUB Go
dira[output]~~                         'output pin
  repeat                               'outer loop
    repeat repfactor
      !OUTA[output]                    'toggle output line
      waitcnt(13_636 +cnt)             'wait to synthesize freq
    waitcnt(delayfact+cnt)             'wait
    repeat repfactor
      !OUTA[output]                    'toggle output line
      waitcnt(12_170 +cnt)             'wait to synthesize freq
    waitcnt(delayfact+cnt)             'wait
    repeat repfactor
      !OUTA[output]                    'toggle output line
      waitcnt(11470 +cnt)              'wait to synthesize freq
    waitcnt(delayfact+cnt)             'wait
    repeat repfactor
      !OUTA[output]                    'toggle output line
      waitcnt(10221 +cnt)              'wait to synthesize freq
    waitcnt(delayfact+cnt)             'wait
    repeat repfactor
      !OUTA[output]                    'toggle output line
      waitcnt(9104 +cnt)               'wait to synthesize freq
    waitcnt(delayfact+cnt)             'wait
```

(continued)

Program 17-2 Code for Generating Audible Tones (*continued*)

```
  repeat repfactor
    !OUTA[output]                    'toggle output line
    waitcnt(8595 +cnt)               'wait to synthesize freq
  waitcnt(delayfact+cnt)             'wait
  repeat repfactor
    !OUTA[output]                    'toggle output line
    waitcnt(7639 +cnt)               'wait to synthesize freq
  waitcnt(delayfact+cnt)             'wait
  repeat repfactor
    !OUTA[output]                    'toggle output line
    waitcnt(6818 +cnt)               'wait to synthesize freq
  waitcnt(delayfact+cnt)             'wait
```

Note *When you run this code within a program, you will notice that the higher frequency notes run for shorter times than the lower frequencies. The reason for this is that the time between the toggles shortens and the notes get done sooner as the frequency increases. (We created eight notes to make sure that we would notice the decrease in the note duration.) This requires that the repeat factor for each note be dependent on the delay between toggles for that specific note. Also note that the duration of each note is not affected by the toggling instruction time or the general program flow (in that each repeating sequence is identical), so if you were in need of an exactly timed sequence, this too would have to be taken into consideration. Think about how you would make each succeeding note play for the same duration.*

We need to be able to read a potentiometer so that we can use the value read to set the eight frequencies we want to play. We will do this with a routine in the Utilities object. The program will use three cogs, assigned as follows:

- **Main cog** Read the potentiometer, read data, and calculate values
- **Cog two** Display LCD information, the tone played, and the potentiometer value
- **Cog three** Play the selected note

The creation of multiple cogs was covered in detail in Part I of the book.

The data we need is added to the program with the following code in a DAT block:

```
DAT
  Tone word 440, 494, 523, 587, 659, 698, 784, 880
```

The data is read from within the program with the following code:

```
PUB GetTone | Tone
  NoteFreq := Tone[toneNumber]
```

Next, we'll put everything together in Program 17-3, which plays eight frequencies based on the position of the potentiometer and displays what is happening on the LCD.

Program 17-3 Generating and Displaying Eight Frequencies with Three Cogs

```
{{ Aug 31 09      Harprit Sandhu
FreqGen8.Spin
Propeller Tool Ver. 1.2.6
Chapter 17 Program 3

This program generates 8 frequencies starting at 440 cycles/sec.
Depends on the position of Pot 1

}}
CON
  _CLKMODE=XTAL1+ PLL2X       'The system clock spec
  _XINFREQ = 5_000_000        'Crystal freq
  Output     = 26             'output line for the speaker
  Repfactor  = 100            'number of times to toggle line

VAR
{RAM assignment for cogs}
  long  Stack1[50]        'this RAM will be assigned to COG1
  long  Stack2[50]        'this RAM will be assigned to COG2

{variables used for reading pot}
  word  PotValue          'value of the pot reading
  long  NoteFreq          'actual frequency of the note
  long  NoteDelay         'delay in cycles

{variables used in displaying LCD in Cog2}
  'none

{variables used in GetTone}
  Byte  toneNumber

OBJ                               'The Objects we will need for our calls
  LCD  : "LCDRoutines4"           'for controlling the LCD
  UTIL : "Utilities"              'for general methods

DAT
  Tone word 440, 494, 523, 587, 659, 698, 784, 880

PUB GO                            'The main method in this program
                                  'the potentiometer is read in this cog
```

(continued)

Program 17-3 Generating and Displaying Eight Frequencies with Three Cogs (*continued*)

```
  cognew (COG_LCD,    @Stack1)    'create COG_TWO for LCD display
  cognew (COG_PLAY,   @Stack2)    'create COG_THREE for toggling output
  repeat                          'start repeat loop
    PotValue:=UTIL.Read3202_0         'get potValue from the utilities
    ToneNumber:=Potvalue/512 +1       'This reduces the value to 0-15

PRI COG_LCD                          'Handles display  to the LCD
  LCD.INITIALIZE_LCD                 'set up the LCD
  REPEAT                             'repeat write to the LCD
     LCD.POSITION (1,1)              'Go to 1st line 1st space
     LCD.PRINT(STRING("PotPos ="))   'Potentiometer position
     LCD.PRINT_DEC(ToneNumber)       'print the tune number
     LCD.SPACE(2)                    'erase over overflows
     LCD.POSITION (2,1)              'Go to 2nd line 1st space
     LCD.PRINT(STRING("FreqOut="))   'frequency label printed
     LCD.PRINT_DEC(NoteFreq)         'print note cps
     LCD.SPACE(2)                    'erase over overflows

PRI COG_PLAY                         'output for speaker
  dira[Output]~~                     'output pin
  repeat                             'outer loop
    repeat RepFactor                 'inner loop and repeat #
      GetTone                        'gets value for NoteFre
      !OUTA[output]                  'toggle output line
      NoteDelay:=12_000_000/(NoteFreq*2)    'calculate the delay
      waitcnt(NoteDelay +cnt)          'wait for delay to synthesize freq
    waitcnt(clkfreq/2+cnt)


PUB GetTone                         'gets the frequency to be played
  NoteFreq := Tone[toneNumber-1]    'looks up position in DAT statements
```

We can generalize this program to generate more frequencies by replacing the routine for determining the frequency from a lookup table like we are using to make a calculation that determines the next frequency. (Each successive frequency is the 12th root of 2 above the previous one.)

We can use this technology to generate whatever frequency we want, limited only by the fact that the wait count has to remain above 381. This sets a limit on how high the frequency generated can be. If higher frequencies are needed, we must use some other technique that does not use waitcnt. For now, we will not go into that.

The wiring diagram for Program 17-3 is shown in Figure 17-1. The speaker module is shown in Figure 17-2.

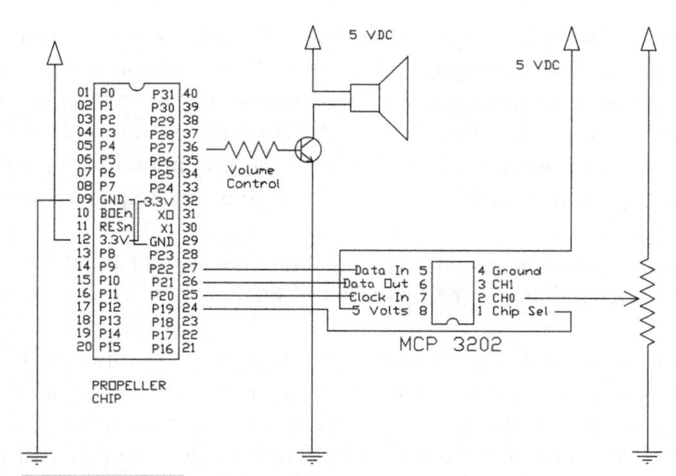

Figure 17-1 Wiring schematic for tone generator

Figure 17-2 Speaker module photo and wiring diagram

Reading Frequencies

There are two basic ways to read a frequency. If we are looking at a low frequency, we can measure how long one wave is and determine the frequency from that. If we have a high frequency, we can count the number of waves we see within a short period of time and determine the frequency from that. In either case, we are assuming that the frequency remains constant and we are assuming that we are looking at essentially square waves. The problems have to do with determining exactly what the effect of the wave shape is on our measurements. To avoid this complication, we are assuming that we are reading square waves. We will read a constant frequency that we will generate.

In general, it is desirable to be able to read any input in less than about 0.01 seconds so that you do not notice a delay in the operation of the program. However, the faster the better.

If the frequency is varying, the situation becomes more complicated. In most cases, the rate at which the frequency is changing is the information of interest rather than the carrier frequency itself. In these cases, the change in frequency is usually a very small percent of the carrier frequency so that we can still read the carrier frequency without much interference from the modulation. Figure 17-3 illustrates the setup for generating and reading the frequency. Figure 17-4 defines the components of the signal being addressed. Program 17-4 lists how the frequency is generated and read.

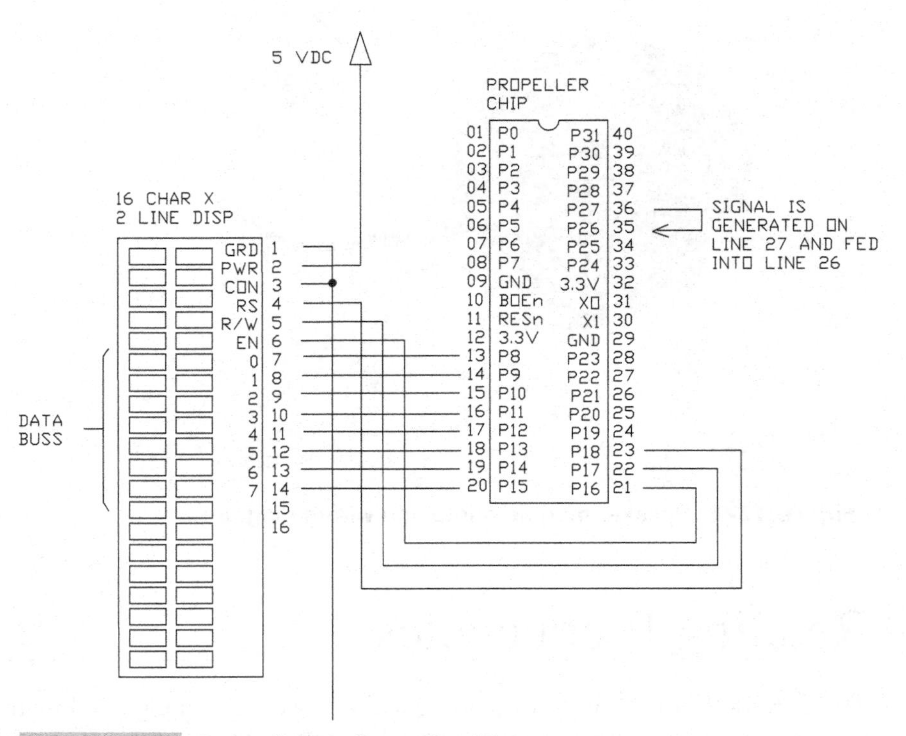

Figure 17-3 Setup for reading a fixed frequency

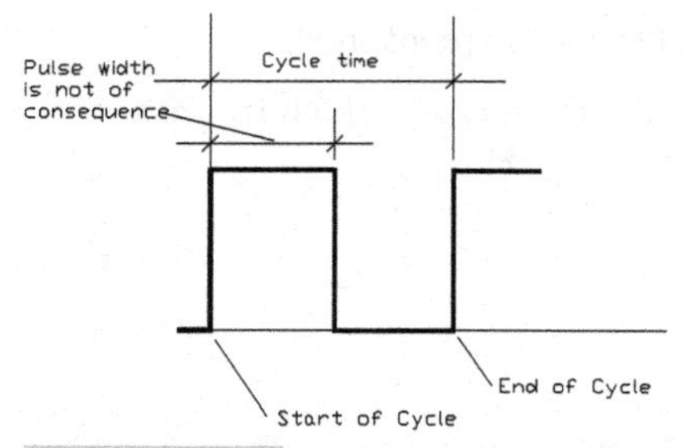

Figure 17-4 Schematic of how cycle time components are identified

Program 17-4 Reading a Generated Frequency

```
{{05 Oct 09     Harprit Sandhu
FreqCounter.Spin
Propeller Tool Ver 1.2.6

The program reads the frequency of a signal on the Input
line and displays it on the LCD. The signal is read as the
number of waves in 1 second.

Three Cogs are used

COG GO reads the frequency
COG COG_LCD displays values on LCD
COG FREQ_GEN generates the frequency

}}
CON
  _CLKMODE=XTAL1+ PLL2X   'The system clock spec
  _XINFREQ = 5_000_000    'crystal
  input    =26            'line for input for what is to be read
  output   =27            'line for output of what is generated

VAR                        'these are the variables
  long  Stack[50]          'For CogOne
  long  Stack1[50]         'For freq_gen
  long  Frequency          'read frequency
  long  startTMRi
  long  StopTMR
  long  elapsed

OBJ                            'These are the methods we will need
  LCD  : "LCDRoutines4"        'for controlling the LCD
  UTIL : "Utilities"           'for general methods
```

(continued)

Program 17-4 Reading a Generated Frequency (*continued*)

```
PUB GO                          'This is the main method in this program
  Cognew (COG_LCD, @Stack)      'create Cog_TWO
  Cognew (FREQ_GEN, @Stack1)    'create freq generator
  dira[input]~
  repeat
    repeat
    while ina[input]==0        'hold at low
    startTMR:=cnt              'start timer
    repeat
    while ina[input]==1        'hold at high
    repeat
    while ina[input]==0        'hold at low
    stopTMR:=cnt               'read timer
    elapsed:=stopTMR-startTMR
    frequency:=10_000_000/elapsed

PRI COG_LCD                       'This is the display, the LCD
  LCD.INITIALIZE_LCD              'initialize up the LCD
  REPEAT                          'Routine to write to the LCD
     LCD.POSITION (1,1)           '
     LCD.PRINT(STRING("FRQ="))    'Frequency
     LCD.PRINT_DEC(Frequency)     'print freq
     LCD.SPACE(2)                 'erase old data

PRI FREQ_GEN
dira[output]~~
   repeat
     outa[output]~~
     waitcnt(clkfreq/5000+cnt)    'change the 5000 to change freq
     outa[output]~
     waitcnt(clkfreq/5000+cnt)    'change the 5000 to change freq
```

Program 17-4 first waits at a low signal and then waits for a rising edge. As soon as the rising edge is detected, the system counter is read and the program starts looking for the signal to go back to a low condition. It then waits for the next rising edge. As soon as the next rising edge is detected, one full wave has gone by, the counter is read again, and the frequency is calculated by dividing the clock frequency by the counts between rising edges. The answer is displayed on the LCD.

The frequency that we are reading is being generated in the freq_gen method. We can change the value (5000) in the two waitcnt lines to change the frequency being generated. Things could be made a little more interesting by reading a potentiometer and making the frequency a function of the potentiometer reading.

18

READING AND CREATING PULSES

Reading Pulse Widths

We are going to investigate the electronic signals generated by a Memsic 2125 accelerometer (as provided by Parallax). This device provides electrical information on two lines that represent the horizontal error/tilt of the sensor in the X and Y directions. That is all I want to say at this time to keep all avenues of investigation open, meaning that we will assume that that is all we know about the sensor as we start our investigation. Let's see what we can find out about the device. This sensor will be used again in Chapter 27 on the self-leveling table.

When we are working with microprocessors, the information of interest to us is often expressed as a series of varying pulse widths. We need to develop the expertise to read these pulse widths. How we read a pulse width depends on how long the pulse width is. There is a limit on how short a pulse width we can read with any microprocessor based on the speed of the processor and the instruction set available to respond to pulses. There is also a kind of a limit on how long a pulse we can read because long pulses may tax our patience, and other devices are better at measuring long intervals than microprocessors. Let's agree that for our purposes anything over 0.25 seconds is a long time. (As a matter of fact, that's a really long time when talking about microprocessors.)

In that we are working in a parallel-processing environment, we can assign one of the cogs to determine how long the pulse width is. Once the width has been determined, it is placed in the shared memory so that all cogs that need to know what the pulse width is can get the information from this one location. If there is a need to know when the information was last refreshed, a flag can be set and cleared when the information is set and read, respectively. One flag will be required for each cog that is going to read the information if it is critical that all the cogs read the information before it is updated.

Other complications have to do with making sure that only the "responsible cog" is allowed to change the pulse width value and that the variable being used to store the pulse width is not altered by any other cog.

Note *Because we do not know when a shared value in the VAR block might be read by one of the cogs, it has to maintain a valid value at all times. This was covered earlier in the book, but is important enough to repeat here.*

In this discussion we are assuming that the pulse width we are interested in either does not vary at all or varies very slowly during the reading process (which is indeed the case for this particular sensor). If rapidly changing values have to be read, that is a whole other problem that we will not discuss in this book. However, you should be able to address the situation after you have successfully performed all the exercises in this book.

Determining the Pulse Width

In Part III of this book, we will be using the Memsic 2125 gravity sensor in an experiment, so we are interested in looking at how this sensor behaves as it is tilted. The information will be used in Chapter 27 to create an intelligent system that maintains a table in a horizontal position in the X and Y directions as the table base is tilted. We know that this sensor provides two pulse widths as its signals, so we will create the software to determine what the characteristics of these pulse widths are, one at a time. The two pieces of information of interest are the cycle time of the signal and the characteristics of the pulse within it. For most sensors, the cycle time remains constant and the pulse width varies, but this might not be the case for this sensor. At this point, we just don't know.

Note *Reading the system clock and using the result in the waitcnt instruction takes a certain amount of time in Spin (381 clock cycles to be exact). Because a minimum for two consecutive reads is needed to read a pulse width, the shortest pulse width that can be read under the best of conditions is 381 clock cycles wide. Because the clock rate is programmable, the width readable also depends on the clock speed specified in our program. We will assume a clock running at our usual 10 MHz for this experiment. If we divide 10,000,000 by 381, we get the highest frequency we can measure in this way. The value is 26,246.7 cps.*

To get a handle on what we are about to undertake, let's note that at 10 MHz, 381 cycles take 381/10,000,000 = 0.0000381 seconds. This equates to a signal running at 26,246.7 Hz. This is a theoretical maximum, meaning that if anything else slows our measurements down, it will get worse—it cannot get better.

For our purpose, the pulses that we are likely to encounter will have pulse widths that vary between about a millionth of a second and about 10 seconds. This range can be laid out in a simple tabular format so we can see what techniques are suitable for what pulse width:

0.000001 seconds	1 microsecond	Too fast. Use I/O compare scheme.
0.00001		Close. Use I/O compare scheme.
0.0001		Over 0.0000381. Can use waitcnt.
0.001	1 millisecond	Can use waitcnt.
0.01		Can use waitcnt.
0.1		Can use waitcnt.
1.0	1 second	Can use waitcnt.
10.0	10 seconds	Can use waitcnt.

For extremely long intervals, the waitcnt instruction has to be used with care because the 32-bit counter overflows every 7.158 minutes (at 10 MHz) and less often at higher frequencies. The value is calculated from:

$$65{,}536 \times 65{,}536/10{,}000{,}000/60 = 7.158 \text{ minutes}$$

We know from the literature provided with the device that each of the two axes on the Memsic provide a pulse width that varies with the tilt of the sensor. The signal is provided at 100 Hz. When the sensor is perfectly horizontal, the signal's duty cycle is 50%. The pulse width increases and decreases by about 25% as an axis is tilted by 90 degrees from horizontal to vertical in either direction. Being good engineers, we will verify this for ourselves by looking at the two signals with an oscilloscope. Make it a habit to look at whatever you are working with, with the many instruments you may have in your shop, to get a preliminary feel for the behavior of the device you are about to employ. The more you know about the device before you start, the greater the chances for your success in using it effectively.

We now have the expertise to use more than one cog to solve a problem. In this program, we will read the pulse widths generated by a Memsic gravity sensor with one cog, display the values on our LCD with another, and output any signal we may need on a third cog. Cog 0, the main cog, will be used to display the length of the pulse (for the X or Y axis), the cycle time, and the frequency of the signal on the two lines of the LCD. Cog 1 will determine the various values and place them in appropriate variables. At the same time, we will look at the signal we are receiving with an oscilloscope so that we can relate what we are seeing to what we are reading with the Propeller chip.

First, we need to read a pulse length with one cog. Let's decide that we are going to measure the high part of the signal. Although this is not always so, the high part is what we mean when we say "pulse width." We can also express the pulse width as a percentage of the total cycle time when that is more appropriate.

The technique we will use to measure the pulse width is as follows:

- We use the system counter as our timer.
- We wait for the signal to go low.
- We then wait for it to go high and then low again to skip a cycle.
- When the signal next goes high, we read the counter immediately.
- We wait for the signal to go low.
- When the signal goes low, we read the counter again immediately.
- We wait for the signal to go high.
- When the signal goes high, we read the counter again immediately.
- We calculate the differences between the counts to get the values we need.
- Each unit is 1/10,000,000 seconds.

The first two high/lows are ignored to make sure we start at a valid edge. (Often sensing one falling edge can be enough to find the next rising edge.) In other words, determining a pulse width is a matter of starting to read the system clock when a signal goes high and adding clock counts to a counter until the signal goes low. If the signal frequency is high and the pulses are short, the time needed to read the data can be quite short. We need to design our program so that we can count a few hundred cycles during the duration of the pulse. A number in the hundreds will allow us to detect a small change in the cycles counted with ease. If we want to have a resolution of 0.5%, we need to read long enough to read over 200 cycles at the point of interest. If we are looking at a range, even the lowest part of the range needs to read 200 cycles. This means that, in general, high frequencies are easier to read and work with than lower frequencies.

As shown in Figure 18-1, the Memsic 2125 provided by Parallax is an eight-pin device. Of these eight pins, two pins are not connected to anything and we will ignore a third (the temperature output). The other five connections are shown in Figure 18-1.

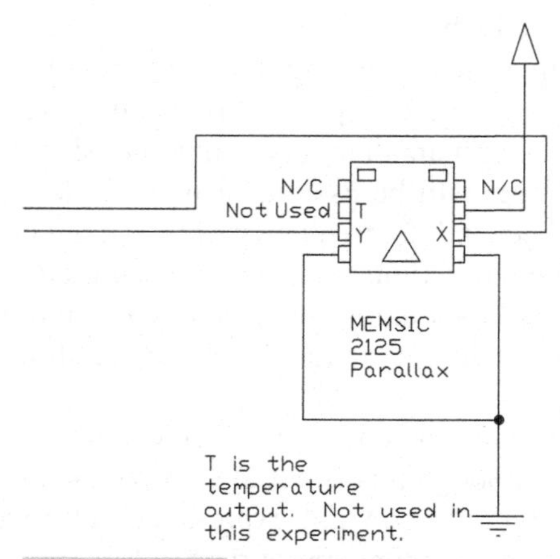

Figure 18-1 Memsic 2125 gravity sensor connections

Figure 18-2 **Memsic 2125 perf board and LCD display. (The frequency being read is 100 Hz.)**

The sensor can be purchased for $25 to $30 (2010) in single quantities. It is mounted on a mini board with pins 0.1 inches on center. We need to mount the device on a board that we can tilt in two directions to see what happens to the signals we are getting. See Figure 18-2 for my breadboard with the sensor mounted on it. See Figure 18-3 for the wiring diagram.

We are going to use the "repeat" and "while" instructions to determine the pulse length in Program 18-1. Refer to the Propeller Manual for a detailed description of these instructions.

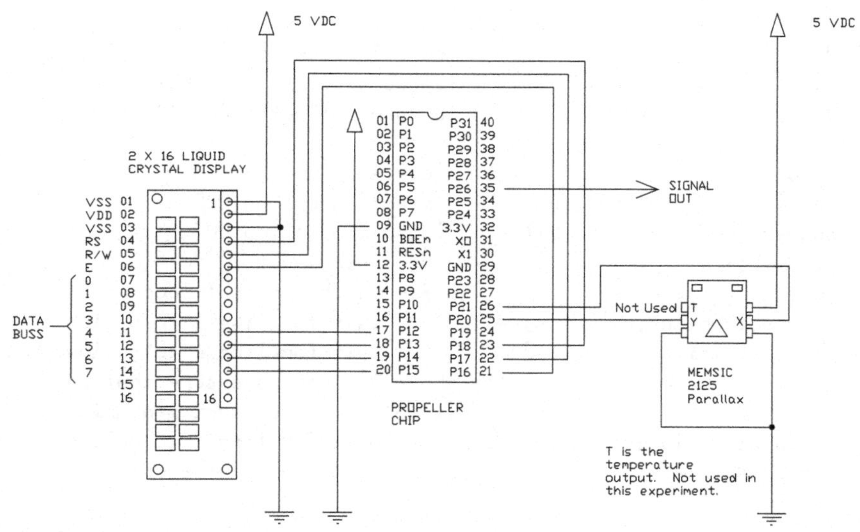

Figure 18-3 **Wiring the Memsic 2125 to a Propeller chip and an LCD**

Program 18-1 Reading One Axis of the Memsic Gravity Sensor: Generating Output to Match at 5× the Read Frequency

```
{{13 Sep 09    Harprit Sandhu
MemsicWidth.spin
Propeller Tool Ver 1.2.6
Chapter 18 Program 1

This program measures the pulse width and cycle time output
from a Memsic 2125 gravity sensor as it is tilted from
the horizontal plane to the vertical plane.

COG_LCD manages the LCD output
COG_0 measures the pulse

}}
CON
  _CLKMODE=XTAL1+ PLL2X        'The system clock spec
  _XINFREQ = 5_000_000         'crystal frequency
  xaxis  = 25                  '
  output = 26
VAR
  long  Stack[55]                'FOR LCD COG
  long  Stack1[55]               'FOR OUTPUT COG
  long  startWave                '
  long  endPulse                 '
  long  endWave                  '
  long  PulseLen                 '
  long  waveLen                  '
  long  frequency                '

OBJ                              'These are the Objects we will need
  LCD  : "LCDRoutines4"          'for controlling the LCD
  UTIL : "Utilities"             'for general methods collection

PUB go                             'Cog_0
  cognew (COG_LCD, @Stack)         'starting up Cog LCD
  cognew (COG_OUT, @Stack1)        'starting up Cog OUT
  DIRA[25]~                        'Make pin input
  repeat                           'Set up the control loop
    repeat while ina[xaxis]==1         'wait for line 1 to go low.
    repeat while ina[xaxis]==0         'wait for line 1 to go low.
                              'the above 2 lines make sure that we see
                              'a full wave when we start measuring
    repeat while ina[xaxis]==1         'wait for line 1 to go low.
    startWave:=CNT                     'read the timer count
```

(continued)

Program 18-1 Reading One Axis of the Memsic Gravity Sensor: Generating Output to Match at 5× the Read Frequency (*continued*)

```
    repeat while ina[xaxis]==0          'wait for line 1 to go low.
    endPulse:=CNT             'read the timer count for second time

    repeat while ina[xaxis]==1          'wait for line 1 to go low.
    endWave:=cnt

    PulseLen:=endPulse-startWave        'figure the pulse
    waveLen:=endWave-startWave          'figure the wave Len
    frequency:=10_000_000/waveLen       'figure the freq
    waitcnt(clkfreq/25+cnt)

PRI COG_LCD                           'This is running in the new cog
  LCD.INITIALIZE_LCD                  'set up the LCD u
  repeat                              'Print to LCD routine
    LCD.POSITION (1,1)                'Position LCD cursor
    LCD.PRINT(String("PL="))          'Pulse
'     LCD.PRINT_DEC((startWave))      'print value
'     LCD.SPACE(2)                    'write over old data
    LCD.PRINT_DEC((pulselen))         'print data
    LCD.SPACE(2)                      'write over old data
    LCD.POSITION (2,1)                'Position LCD cursor
    LCD.PRINT(String("WL="))          'Cycle time
    LCD.PRINT_DEC((wavelen))          'print value
    LCD.SPACE(2)                      'write over old data
    LCD.POSITION (2,11)               'Position LCD cursor
    LCD.PRINT(String("FR="))          'Frequency
    LCD.PRINT_DEC((frequency))        'print value
    LCD.SPACE(4)                      'write over old data

PRI COG_OUT                           'output signal generation.
  dira[output]~~                      'Set pin direction as output
  repeat                              'loop
    outa[output]~~                    'make line high
    waitcnt(pulselen/10+cnt)          'Create hi part of wave
    outa[output]~                     'Make line low
    waitcnt(100_000-pulselen+cnt)'Create rest of wave
```

If we wanted to know the frequency of the signal we were reading, we would measure both the high and low parts of the signal and add them together to get the cycle time for one wave cycle. It is interesting to see how much the signal pulse width varies as the sensor is tilted in the read direction. Run the experiment and note your results.

In Program 18-1, we read one axis of the Memsic. Later on in Part III of the book, we expand on this program to read both axes as a part of a more sophisticated program that we will write for our auto-leveling experiment.

Pulse Width Creation

Now that we have determined what the incoming pulse looks like we are ready to create a similar pulse on an output line. We will create the pulse on line P27 at five times the frequency of the incoming pulse so that the two pulses are easily differentiated. Two items need to be specified to define a pulse width: the frequency of the signal and the width/range of the pulse width within the signal. If the range is not specified, we can assume it goes from 0% to 100% of the cycle time.

Let's create a square waveform at 500 Hz (the incoming pulse was at 100 Hz) in which we match the pulse width (%) of the incoming signal. This frequency is five times the frequency we read at the Memsic sensor so that we can readily see the difference between the two signals. Figure 18-4 shows what we are trying to accomplish.

Note *Both signals can be displayed on a oscilloscope at the same time.*

At 100 Hz and a frequency clock of 10,000,000 Hz, each cycle time is 100,000 cycles long. Within this wave we have to vary the pulse width so that it follows what we are reading from the Memsic sensor. We want to make it as close as we can, and we should be able to make it pretty close.

We are speeding things up by 5×. Therefore, the new pulse length will be 1/5th of whatever we are reading.

The code shown in Program 18-2 will run in an independent cog.

Program 18-2 Code Segment to Generate 5× Output

```
PRI COG_OUT                           'output signal generation.
  dira[output]~~                      'Set pin direction as output
  repeat                              'loop
    outa[output]~~                    'make line high
    waitcnt(pulselen/5+cnt)           'Create hi part of wave
    outa[output]~                     'Make line low
    waitcnt(200_000-pulselen+cnt)     'Create rest of wave
```

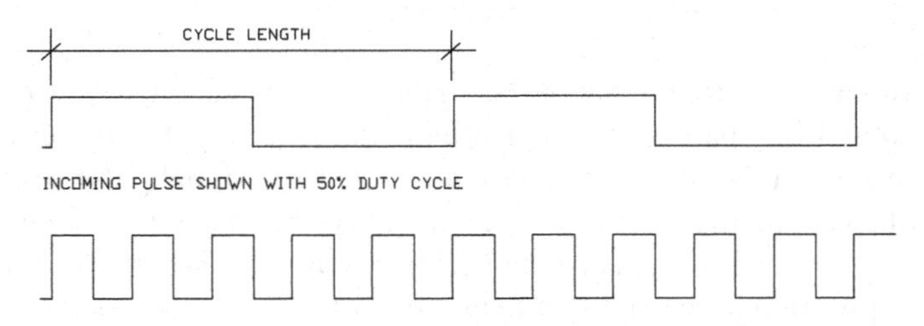

Figure 18-4 Waveforms of read and created pulse widths

This code is incorporated at the bottom of Program 18-1 to complete the generation of the outgoing signal. A complete relisting of the entire program with a potentiometer variable added to create the initial frequency is given in Program 18-3.

Program 18-3 Code for Original and 5× Output

```
{{05 Oct 09     Harprit Sandhu
FreqInOut.Spin
Propeller Tool Ver 1.2.6
Chapter 18 Program 3

The program reads the frequency of a signal on the Input
line and displays it on the LCD. The signal is read as the
number of waves in 1 second. Controlled by Pot 1

Four Cogs are used

COG GO reads the frequency
COG COG_LCD displays values on LCD
COG FREQ_GEN generates the frequency
COG COG_OUT generates the 5 freq

}}
CON
  _CLKMODE=XTAL1+ PLL2X    'The system clock spec
  _XINFREQ = 5_000_000     'crystal
  freq5    =25             '5X freq line
  input    =26             'line for input for what is to be read
  output   =27             'line for output of what is generated

VAR                         'these are the variables
  long  Stack[50]           'For CogOne
  long  Stack1[50]          'For freq_gen
  long  Stack2[50]          'For freq_out
  long  Freq                'read frequency
  long  startTMR
  long  StopTMR
  long  elapsed
  long  pot

OBJ                             'These are the methods we will need
  LCD  : "LCDRoutines4"         'for controlling the LCD
  UTIL : "Utilities"            'for general methods

PUB GO                          'This the main method in this program
  Cognew (COG_LCD, @Stack)      'create Cog_TWO
  Cognew (FREQ_GEN, @Stack1)    'create freq generator
```

(continued)

Program 18-3 Code for Original and 5× Output (*continued*)

```
  Cognew (COG_OUT, @Stack2)    'create freq out
  dira[input]~
  repeat
    repeat
    while ina[input]==0         'hold at low
    startTMR:=cnt               'start timer
    repeat
    while ina[input]==1         'hold at high
    repeat
    while ina[input]==0         'hold at low
    stopTMR:=cnt                'read timer
    elapsed:=stopTMR-startTMR
    freq:=10_000_000/elapsed
    pot:=util.read3202_0         'read Pot 1

PRI COG_LCD                      'This is the display, the LCD
  LCD.INITIALIZE_LCD             'initialize up the LCD
  REPEAT                         'Routine to write to the LCD
     LCD.POSITION (1,1)          '
     LCD.PRINT(STRING("FRQ="))   'Frequency
     LCD.PRINT_DEC(Freq)         'print freq
     LCD.SPACE(2)                'erase old data
     LCD.POSITION (2,1)          '
     LCD.PRINT(STRING("POT="))   'Frequency
     LCD.PRINT_DEC(Pot)          'print Pot reading
     LCD.SPACE(2)                'erase old data

PRI FREQ_GEN                     'Freq generation based on pot reading
dira[output]~~                   'set output line
   pot:=1000                     'arbitrary non zero value
   repeat                        'loop
     outa[output]~~              'make line high
     waitcnt(clkfreq/pot+cnt)    'wait based on pot
     outa[output]~               'make line low
     waitcnt(clkfreq/pot+cnt)    'wait based on pot

PRI COG_OUT                         'output signal generation.
  dira[freq5]~~                     'Set pin direction as output
  freq:=1000                        'arbitrary non zero value
  repeat                            'loop
    outa[freq5]~~                   'make line high
    waitcnt(clkfreq/freq/5+cnt)     'wait 1/5 time for 5X freq
    outa[freq5]~                    'Make line low
    waitcnt(clkfreq/freq/5+cnt)     'wait 1/5 time for 5X freq
```

Part III

THE PROJECTS: USING WHAT WAS LEARNED TO BUILD THE PROJECTS

Now that we understand the basics of manipulating the bits and bytes in the Propeller let us use the knowledge we have gained to create some real world applications. Here we will concentrate on developing the type of skills that robot builders would find useful for running various types of motors. The principles used to run motors are universal and can be adapted to serve any device that needs to be controlled with a microcontroller. Pay particular attention to controlling the DC motor with an encoder; it represents all systems with feedback and is the prototypical control application.

19

SEVEN-SEGMENT DISPLAYS: DISPLAYING NUMBERS WITH SEVEN-SEGMENT LED DISPLAYS

This chapter is about doing a lot of things so quickly that it looks like they are all being done at the same time.

Sometimes we need to display a number consisting of a few digits as a part of the project we are undertaking. The easiest way to do this is to use as many seven-segment LED displays as necessary to get the number displayed. Although it would be possible to light up all the segments necessary to display a number at the same time, the usual practice is to light them one at a time in rapid succession. If this is done right, the human eye/brain cannot tell the difference, and a lot less power and fewer lines are needed to control the display. Instead of two lines for each segment in each number, we can get by with seven or eight lines plus one line for each numeral. Eight lines will be needed if we need to control the decimal points also. The scheme requires considerably less current because we will be lighting only one segment at a time. We are not going to need the 7404 buffers in this program because we will be lighting one segment at a time only and we know the Propeller can handle that without overloading its circuitry. Figure 19-1 illustrates a typical seven-segment display using four seven segment devices. The diagram for identifying the connections to the various segments for the connecting pins of each display is provided in Figure 19-2.

Note *A typical four-character seven-segment display is good for 0 to 9999.*

Program 19-1 uses a potentiometer to vary the time between the illumination of the number again and again. This is a brute-force program that turns the segments on one after another, and the delay between successive illuminations is controlled with the potentiometer. The potentiometer determines how may times each number will be displayed before going on to the next number. Although some segments need fewer

Figure 19-1 Typical seven-segment displays

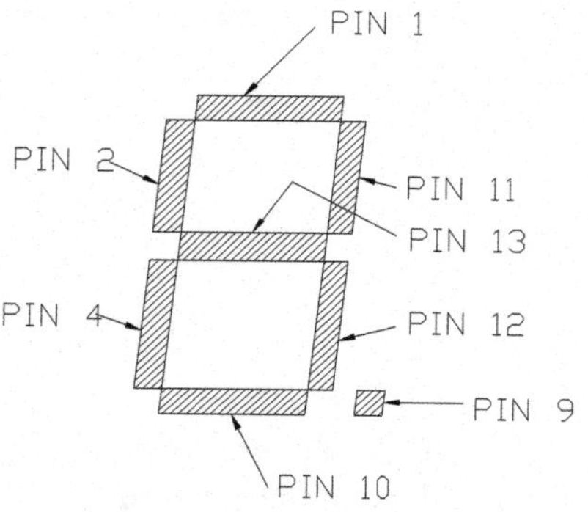

Figure 19-2 Pin assignment for a common anode seven-segment display (face view, as seen from above). These are the pins as assigned on the 16-pin device.

illuminations than others (if a segment is smaller it needs less current for the same brightness), each illumination in this particular program sequence addresses all eight segments identically so that the time taken will be the same for each segment displayed.

In the program we are using, the segments are connected to the Propeller as shown in Figure 19-3. The actual layout that we used is shown in Figure 19-4.

Note *These are the pin connections to the Propeller, using same pins as the LCD uses.*

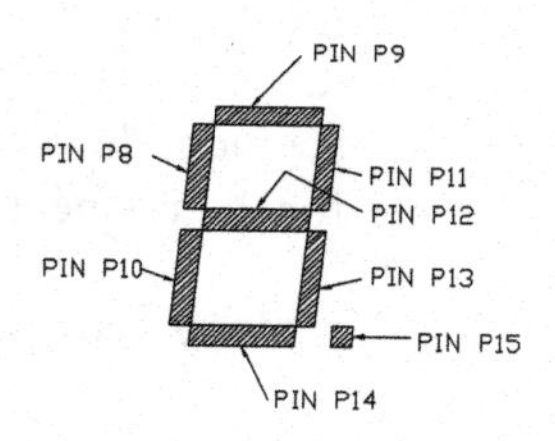

Figure 19-3 Actual segment connections to the Propeller

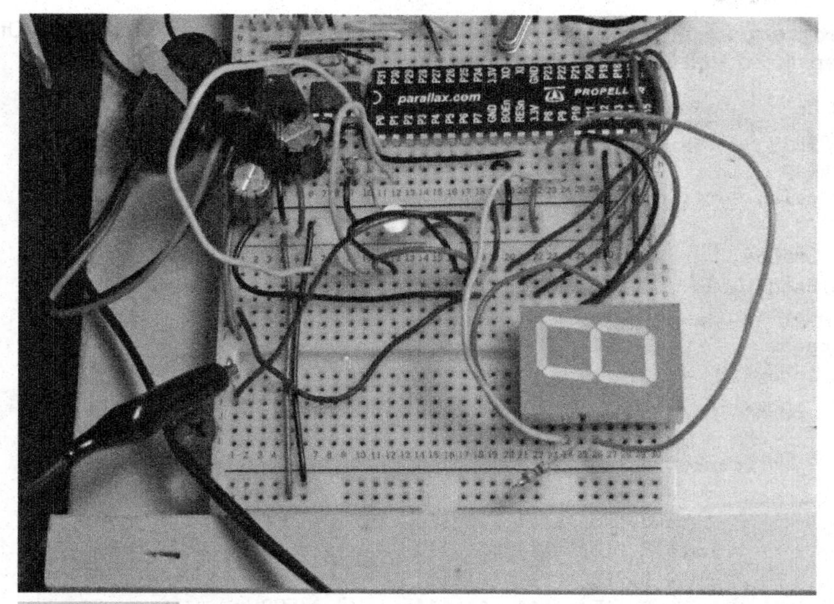

Figure 19-4 Positioning the seven-segment display on the bread board

Program 19-1 Displaying All the Segments on One Display, One after the Other, Rapidly So That They Look Like They Are All on Simultaneously

```
{{28 Sep 09  Harprit Sandhu
7SegDisplay.spin
Propeller Tool Ver 1.2.6
Chapter 19 Program 1

Program controls the segments on one 7 segment display
Pot is used to change the delay time
Place Segment Display on perf board as shown in Pictures.

}}
CON
  _CLKMODE=XTAL1+ PLL2X        'The system clock spec
  _XINFREQ     =5_000_000      '10 MHz

VAR                           'these are the variables we will use.
  long stack1[35]             'space for 7 Seg driver
  word delay                  'duration for each number
  byte index                  'numbering index

OBJ                           'These are the methods we will need
  UTIL : "Utilities"          'for general methods
```

(continued)

Program 19-1 Displaying All the Segments on One Display, One after the Other, Rapidly So That They Look Like They Are All on Simultaneously (*continued*)

```
PUB Go                             'main Cog
  Cognew(SevSeg,@Stack1)     'start new Cog for Segments driver
  repeat                           'this main Cog's main loop
    delay:=util.read3202_0/16

PRI SevSeg
  dira[8..15]~~
  index:=0
  repeat
    index+=1
    'index:=8     'use this line to freeze index at 8 to see all segs lit
        1:
          repeat delay
            outa[8..15]:=%01_111_111        'These numbers build
            outa[8..15]:=%11_111_111        'up the numbers shown
            outa[8..15]:=%11_011_111        'a segment at a time.
            outa[8..15]:=%11_111_111        'One number at a time.
            outa[8..15]:=%11_111_111        'a ZERO turns a segment on
            outa[8..15]:=%11_111_111
            outa[8..15]:=%11_111_111
            outa[8..15]:=%11_111_110
        2:
          repeat delay
            outa[8..15]:=%11_111_111
            outa[8..15]:=%10_111_111
            outa[8..15]:=%11_011_111
            outa[8..15]:=%11_101_111
            outa[8..15]:=%11_110_111
            outa[8..15]:=%11_111_111
            outa[8..15]:=%11_111_101
            outa[8..15]:=%11_111_110
        3:
          repeat delay
            outa[8..15]:=%11_111_111
            outa[8..15]:=%10_111_111
            outa[8..15]:=%11_111_111
            outa[8..15]:=%11_101_111
            outa[8..15]:=%11_110_111
            outa[8..15]:=%11_111_011
            outa[8..15]:=%11_111_101
            outa[8..15]:=%11_111_110
```

(*continued*)

Program 19-1 Displaying All the Segments on One Display, One after the Other, Rapidly So That They Look Like They Are All on Simultaneously (*continued*)

```
4:
  repeat delay
    outa[8..15]:=%01_111_111
    outa[8..15]:=%11_111_111
    outa[8..15]:=%11_101_111
    outa[8..15]:=%11_110_111
    outa[8..15]:=%11_111_011
    outa[8..15]:=%11_111_111
    outa[8..15]:=%11_111_111
    outa[8..15]:=%11_111_110
5:
  repeat delay
    outa[8..15]:=%01_111_111
    outa[8..15]:=%10_111_111
    outa[8..15]:=%11_110_111
    outa[8..15]:=%11_111_011
    outa[8..15]:=%11_111_101
    outa[8..15]:=%11_111_111
    outa[8..15]:=%11_111_111
    outa[8..15]:=%11_111_110
6:
  repeat delay
    outa[8..15]:=%01_111_111
    outa[8..15]:=%10_111_111
    outa[8..15]:=%11_110_111
    outa[8..15]:=%11_011_111
    outa[8..15]:=%11_111_011
    outa[8..15]:=%11_111_101
    outa[8..15]:=%11_111_111
    outa[8..15]:=%11_111_111
7:
  repeat delay
    outa[8..15]:=%11_111_111
    outa[8..15]:=%10_111_111
    outa[8..15]:=%11_101_111
    outa[8..15]:=%11_111_011
    outa[8..15]:=%11_111_111
    outa[8..15]:=%11_111_111
    outa[8..15]:=%11_111_111
    outa[8..15]:=%11_111_111
```

(*continued*)

Program 19-1 Displaying All the Segments on One Display, One after the Other, Rapidly So That They Look Like They Are All on Simultaneously (*continued*)

```
    8:
      repeat delay
        outa[8..15]:=%01_111_111
        outa[8..15]:=%10_111_111
        outa[8..15]:=%11_011_111
        outa[8..15]:=%11_101_111
        outa[8..15]:=%11_110_111
        outa[8..15]:=%11_111_011
        outa[8..15]:=%11_111_101
        outa[8..15]:=%11_111_111
    9:
      repeat delay
        outa[8..15]:=%01_111_111
        outa[8..15]:=%10_111_111
        outa[8..15]:=%11_101_111
        outa[8..15]:=%11_110_111
        outa[8..15]:=%11_111_011
        outa[8..15]:=%11_111_101
        outa[8..15]:=%11_111_111
        outa[8..15]:=%11_111_111
    10:
      repeat delay
        outa[8..15]:=%01_111_111
        outa[8..15]:=%10_111_111
        outa[8..15]:=%11_011_111
        outa[8..15]:=%11_101_111
        outa[8..15]:=%11_111_111
        outa[8..15]:=%11_111_011
        outa[8..15]:=%11_111_101
        outa[8..15]:=%11_111_111
    11:
      repeat delay*20   'blanks display for a while
        outa[8..15]:=%11_111_111

if index>11
  index:=0
```

One display can be placed comfortably on the education board after the LCD display is removed. We will use the same lines as were used to control the LCD to control the seven-segment display (7SD). This will keep wiring disruption to a minimum when we want to return to the use of the LCD for further experimentation.

The segments are lighted one segment at a time, as shown in Program 19-1. Each digit is displayed *delay* number of times. The displaying sequence is so fast that we do not notice the blinking of the segments.

Because only one segment is lit at any one time, the Propeller electronics are not overloaded and we do not have to go through the buffers for more power.

Here are the rules we followed:

- Only one task in each cog.
- Only one segment can be on at any one time.
- Run the segment changes as fast as possible.
- Use the common anode design of the 7SD to our advantage so the Propeller is used as a sink rather than a source of power.

If we were using more than one display, we would wire all the segments in parallel and connect them to the Propeller one at a time by selecting the common anode connection of each one in a round-robin fashion. This is shown schematically in Figure 19-5.

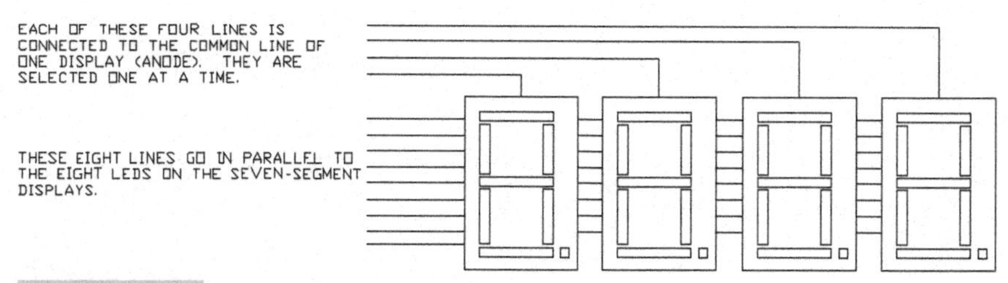

Figure 19-5 Using four displays

20

THE METRONOMES

This chapter is about creating accurate, controlled pulse sequences.

Ms. Music, the local high school music teacher, did not get the funding for the metronomes she needs for her class. The enterprising local principal has asked your electronics instructor to see if he can get your class to make 25 low-cost electronic metronomes for the music students.

Creating an electronic metronome is an exercise in producing discrete, accurate time intervals that are controlled from a potentiometer.

On the metronomes we create, the counts/minute will be controlled by a potentiometer, they will be shown on the display, and they will be annunciated on a speaker. Although the output needed from a metronome is rather limited (40 to 208 ticks per minute is the standard), our project will be able to provide any count desired, from a count every few seconds to a few thousand counts per second, when we use appropriate values in our program. All we need to do is to change the software (and that, in a nutshell, is the power of devices that you make yourself).

To implement this in a multiprocessor parallel-processing scenario, we will use four cogs:

- **Cog0** blinks an LED at a fixed rate (60 cycles per minute) to indicate that the system is on. We use a rate of exactly one-sixtieth of a second for the LED to serve as a comparative standard for the metronome operations. When the metronome is set for 60 beats per minute, it should match the blinking of the LED exactly.
- **Cog1** manages the LCD.
- **Cog2** actuates the speaker coil to create a click.
- **Cog3** reads the potentiometer and calculates the needed values.

The complete program for the metronome is provided in Program 20-1. This program illustrates how easy it is to compartmentalize the four tasks that need to be undertaken by assigning them to individual cogs and how effortless it is for the cogs to share the information they need to make it all work.

The wiring for a metronome is provided in Figure 20-1.

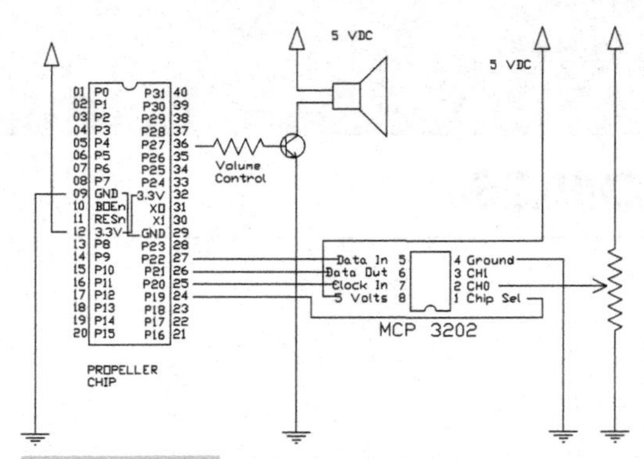

Figure 20-1 Wiring for the metronome

Program 20-1 Electronic Metronome

```
{{29 Sep 09   Harprit Sandhu
Metronom1.Spin
Propeller Tool Ver 1.2.6
Chapter 20 Prog 1

Creating pulses
An electronic metronome.
Standard is 40 to 208 BPM (beats per minute)

4 Cogs are used
Cog0 GO, toggles power line once a sec
Cog1 LCD, displays values on LCD
Cog2 Counts, toggles the speaker
Cog3 Calc, calculates the beats

}}
CON
  _CLKMODE=XTAL1+ PLL2X     'The system clock spec
  _XINFREQ   = 5_000_000    'crystal rate

  Clickline    = 25       'Clicks speaker
  LED2Line     = 26       'Toggles the 2nd LED
  ledLine      = 27       'the ON led, 1 sec cycle

VAR                       'these are the variables we use.
  long  Stack [40]        'for lcd
  long  Stack2[40]        'for counts
  long  Stack3[40]        'for spkr
  long  Stack4[40]        'for click
  long  Pot               'for potentiometer reading
  long  bpm               'beats per minute
```

(continued)

Program 20-1 Electronic Metronome (*continued*)

```
OBJ                          'These are the methods we need
  LCD  : "LCDRoutines4"      'for controlling the LCD
  UTIL : "Utilities"         'for general methods, read potentiometer

PUB GO                               'This the main method in this program
  cognew (lcd_COG,  @Stack)          'create COG for LCD
  cognew (Counts_COG, @Stack2)       'create COG for speaker
  cognew (Calc_COG, @Stack3)         'create COG for calculations
  dira[ledLine]~~                    'Power indicator
  repeat                             'The main cog blinks an LED only
    outa[ledLine]~                   'These lines just blink an LED
    waitcnt(clkfreq/2+cnt)           'as an ON indicator as a one
    outa[ledLine]~~                  'second reference
    waitcnt(clkfreq/2+cnt)           'line 27

PRI lcd_COG                          'This is the display to the LCD
  LCD.INITIALIZE_LCD                 'set up the LCD
  REPEAT                             'Routine to write to the LCD
    LCD.POSITION (1,1)               'Position cursor
    LCD.PRINT(STRING("POT VAL="))      'identify signal
    LCD.PRINT_DEC(pot)               'print the Pot reading
    LCD.SPACE(4)                     'write over old data
    LCD.POSITION (2,1)               'Position cursor
    LCD.PRINT(STRING("BEATS/M="))      'identify signal
    LCD.PRINT_DEC(bpm)               'print the beats/min
    LCD.SPACE(4)                     'write over old data
                                     '
PRI Counts_COG                       'This COG toggles the speaker
  dira[LED2Line]~~                   '26
  bpm:=60                            'this initializes the value
  repeat                             'repeat loop
    outa[LED2Line]~~                 'toggle second LED
    waitcnt(clkfreq/(bpm/30)+cnt)    'wait time
    Click_Speaker                    'calls method for this
    outa[LED2Line]~                  'turn it off
    waitcnt(clkfreq/(bpm/30)+cnt)    'wait time

PRI Calc_COG                         'This COG calculates the beats
  pot:=10                            'this initializes the value
  bpm:=60                            'this initializes the value
  repeat                             'repeat loop
    pot:=util.Read3202_0             'read the potentiometer
    bpm:=40+(10*pot)*(208-40)/40950    'calculate beats from pot reading

PRI Click_speaker                    'makes the click sound
  dira[Clickline]~~                  'line 25
  outa[Clickline]~~                  'turn on
  outa[Clickline]~                   'turn off
```

21

UNDERSTANDING A 16-CHARACTER-BY-2-LINE LCD DISPLAY

In general, this chapter is about interfacing an external device to the Propeller using a liquid crystal display (LCD) as the target device. Specifically, this chapter is about attaching various LCDs to your projects. Almost all the LCDs on the market with built-in controllers use the scheme we are about to investigate to communicate with them. Displays that go from 8 characters on one line to 40 characters on each of four lines are available. All these displays can be addressed with the techniques discussed in this chapter.

We need some way to see what is going on in our programs as we develop them. We will focus our efforts on a 16-character-by-2-line LCD (see Figure 21-1). These LCDs can be purchased for between $4.50 and $6 on the Internet. They provide an inexpensive

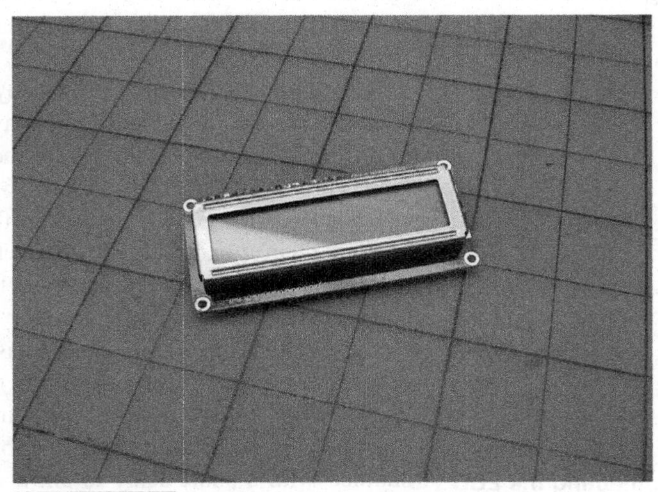

Figure 21-1 A 16-character-by-2-line LCD module

way of adding a display to our projects, and the 32 characters they display provide enough information for most simple experiments. We will develop the software to control this LCD in the following experiments, and we will use this LCD as the display for all our experiments. This being the case, we need a comprehensive set of routines to allow us to have complete control of this device. The basic routines will be developed as we proceed. You can add to them as your proficiency and needs increase.

Note *A one- or two-line LCD can add a tremendous amount of utility to a project.*

Our goal in this experiment is to get the LCD up and running from its most primitive first instructions. A number of very specific instructions have to be sent to the LCD to start it up. These are specified in the startup instruction set for the LCD specified in the manual for the Hitachi 44780 controller. This Integrated Circuit controls the internal workings of the LCD and almost all similar LCDs on the market. Our particular LCD can be controlled by sending data to it either 4 bits at a time or 8 bits at a time. We will cover both techniques, with the 8-bit protocol first.

It is useful to have the full 40+ page data sheet on hand when you are doing anything more than sending text to the LCD. The Hitachi 44780 data sheet can be downloaded from the Internet by searching "Hitachi 44780" and looking for a suitable PDF version of the datasheet. Datasheets can also be downloaded from the websites of most electronic component suppliers like Digikey, Mouser, and Avnet.

8-Bit Mode

The circuitry needed to connect the LCD to the Propeller chip is shown in Figure 21-2. Although the LCD can be written to 4 bits at a time or all 8 bits at one time, the diagram shows all eight lines connected to the data bus so that we can experiment with either 4 or 8 bits. After all is said and done, we will use the 4-bit mode so that we can free up four lines on the Propeller.

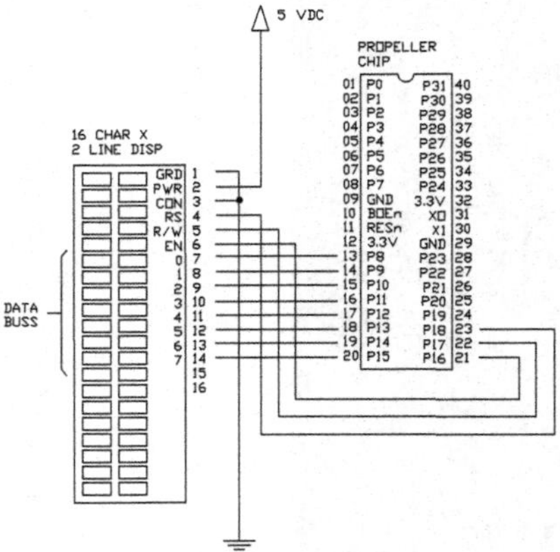

Figure 21-2 **Connecting the LCD to the Propeller chip (8-bit mode)**

It is possible to write to the LCD by sending it the characters we want it to display as well as to read from the LCD to determine what is in the current LCD memory. These 16×2 LCDs can display only 32 characters, but they have enough memory to store 80 characters. The display can be scrolled left and right to see the hidden characters. If you are never going to read from the LCD (and reading is seldom necessary, except to read the Busy bit), you can tie the R/W line low and save one line on the Propeller for other uses. (The busy bit is then handled with a short delay to allow it to turn off.) Because using 4 bits for communications saves another 4 lines, the LCD can be controlled with a minimum for six lines. We will connect all 11 lines for now to allow uninhibited experimentation.

The Spin language does not have a print command as such built into it. It is possible to send a target device one character at a time, but strings are not permitted. We will first create a command to send the LCD one character at a time and then build up a routine that will allow us to send a string. Here are the other commands needed:

- Initialize the LCD.
- Clear the LCD.
- Position the cursor at a given line and horizontal position.
- Send character (already mentioned).
- Send command.
- Print a string.
- Print a decimal value.
- Print a hexadecimal value.
- Print a binary value.
- Print a space (or spaces) to overwrite old information.

We also need to be able to send the LCD certain nonprinting instructions, and this is done with the "send command" instruction mentioned in the preceding list.

Let's first build up the methods that control the LCD; they use the Register Select line, the Read/Write line, and the Enable line.

We set the LCD up for initial use by sending it some special instructions that specify the number of bits, number of lines, font size, cursor condition, and other factors. These instructions and their proper uses are described in the Hitachi manual.

INITIALIZING THE LCD

Before we can do anything with it, we have to initialize the liquid crystal display. The commands we have to send to the LCD are specified in the Hitachi manual. Setting this up consists of sending the commands shown in Program 21-1.

Program 21-1 Code Segment for Initializing the LCD

```
PRI INITIALIZE_LCD                'The addresses and data used here are
  waitcnt(150_000+cnt)            'specified in the Hitachi data sheet for the
                                  'display. YOU MUST CHECK THE DATA SHEET.
  OUTA[RegSelect] := 0            'These lines are specified to write
  OUTA[ReadWrite] := 0            'the initial set up bits for the LCD
  OUTA[Enable]    := 0            'See Hitachi HD44780 data sheet

  SEND_INSTRUCTION (%0011_1000)      'Sets DL=8 bits, N=2 lines, F=5x7 font
  waitcnt(50_000+cnt)'               'wait for instruction to execute
  SEND_INSTRUCTION (%0000_1111)      'Display on, Cursor on, Blink on
  waitcnt(12_000+cnt)                'wait for instruction to execute
  SEND_INSTRUCTION (%0000_0110)      'Move Cursor, Do not shift display
  waitcnt(12_000+cnt)                'wait for instruction to execute
```

At the end of these instructions, you are ready to communicate with the LCD.

CLEARING THE LCD

The next command we need is one that clears the LCD no matter what it may have in its display. This will be the first instruction we send the LCD after a program startup and whenever we want to clear the display. The instructions for effecting an "LCD clear" are shown in Program 21-2.

Program 21-2 Code Segment to Clear the LCD

```
PUB CLEAR                                   'Clear the LCD display and go home
  SEND_INSTRUCTION (%0000_0001)             'to the 1,1 position
```

POSITIONING THE CURSOR ON A GIVEN LINE AND AT A GIVEN HORIZONTAL POSITION

Now that we know how to initialize the LCD and how to clear it, we are ready to position the cursor at the desired line and the desired horizontal position so that the first character can be displayed where we want. We have to pass a line number and the horizontal position to this method. We define these two variables at the top of the method, as is standard practice in Spin. These are local variables. The code for positioning the cursor is shown in Program 21-3.

Program 21-3 Code Segment to Position the Cursor

```
PUB POSITION (LINE_NUMBER, HOR_POSITION) | CHAR_LOCATION
      'Position the cursor at location
      'HOR_POSITION : Horizontal Position : 1 to 16
      'LINE_NUMBER : Line Number : 1 or 2
CHAR_LOCATION := (HOR_POSITION-1) * 64    'figure location
CHAR_LOCATION += (LINE_NUMBER-1) + 128    'figure location
SEND_INSTRUCTION (CHAR_LOCATION)          'send instr to pos crsr
```

Note that character location is calculated twice. The first line determines the horizontal position for the output. The next lines add 128 to the value calculated for line 1, or 129 if it is calculated for line 2. This is as specified in the Hitachi manual.

The Busy Bit There is a bit in the LCD that is set to 1 when the LCD is busy and cannot accept any more data from outside. If we want to operate as rapidly as possible, we need to wait until this bit is set to zero and then send the next instruction immediately. The Read/Write bit and the Register Select bit have to be set as shown to access the Busy bit. The code to read the Busy bit is provided in Program 21-4.

Program 21-4 Code Segment to Check the Busy Bit

```
PRI CHECK_BUSY | BUSY_BIT                           'routine to check busy bit
  OUTA[ReadWrite] := 1                              'Set to read the busy bit
  OUTA[RegSelect] := 0                              'Set to read the busy bit
  DIRA[DataBit7..DataBit0] := %0000_0000            'Set the port to be an input
  REPEAT                                            'Keep doing it till clear
    OUTA[Enable]  := 1                    'set to 1,  to toggle H>L this bit
    BUSY_BIT := INA[DataBit7]             'the busy bit is bit 7
    OUTA[Enable]  := 0                    'enable bit low = H>L toggle
  WHILE (BUSY_BIT == 1)                   'do it as long as the busy bit is 1
  DIRA[DataBit7..DataBit0] := %1111_1111  'done,  port bck to output
```

SEND CHARACTER

Once we have positioned the cursor, we are ready to send the alphanumeric data we want to display to the LCD. The LCD is prepared for data reception by setting the three control lines as shown in Program 21-5. Each time the Enable line is toggled, the data on the bus is transferred to the LCD. The data must be sent a byte at a time.

Program 21-5 Code Segment to Send a Single Character to the LCD

```
PRI SEND_CHAR (DISPLAY_DATA) 'set up for writing to the display
  CHECK_BUSY                 'wait for busy bit to clear before sending
  OUTA[ReadWrite] := 0                           'Set up to read busy bit
  OUTA[RegSelect] := 1                           'Set up to read busy bit
  OUTA[Enable]    := 1                           'Set up to toggle bit H>L
  OUTA[DataBit7..DataBit0] := DISPLAY_DATA       'Ready to SEND data in
  OUTA[Enable]    := 0                           'Toggle the bit H>L
```

SEND INSTRUCTION

We also need to be able to send the LCD non-alphanumeric instructions. This is similar to the character routine and is shown in Program 21-6.

Program 21-6 Code Segment to Send an Instruction to the LCD

```
PRI SEND_INSTRUCTION (DISPLAY_DATA)       'set up for writing instructions
  CHECK_BUSY                              'wait for busy bit to clear
  OUTA[ReadWrite] := 0                    'Set up to read busy bit
  OUTA[RegSelect] := 0                    'Set up to read busy bit
  OUTA[Enable]    := 1                           'Set up to toggle bit H>L
  OUTA[DataBit7..DataBit0] := DISPLAY_DATA       'Ready to READ data in
  OUTA[Enable]    := 0                           'Toggle the bit H>L
```

At this time, all these methods are defined as being private to this object. Later on we will make them public, and any method in another procedure will be able to call and use them.

When we combine all the preceding code into a program, we get the listing in Program 21-7.

Program 21-7 Minimal Program to Send Characters to the LCD

```
{{11 Sep 09      Harprit Sandhu
LCDminimal.spin
Propeller Tool Ver 1.2.6
Chapter 21 Program 7

PROGRAM TO BEGIN USING THE LCD

A minimal LCD implementation to allow us to use the LCD in our
experiments immediately.

This program is an absolutely minimal implementation to
make the LCD usable. You can work on improving it.
```

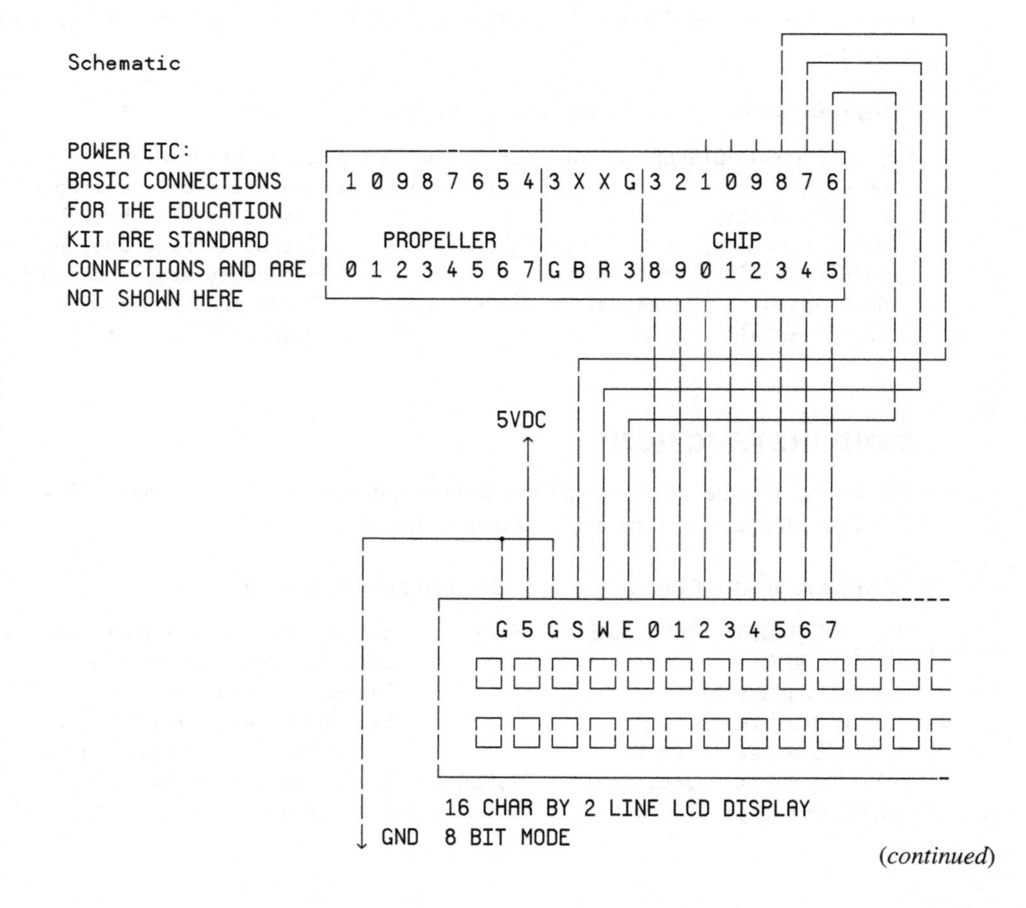

(continued)

Program 21-7 Minimal Program to Send Characters to the LCD (*continued*)

```
Revisions



}}

CON
{{
Pin assignments are assigned as constants because the pins are fixed.
These numbers reflect the actual wiring on the board between the Propeller
and the 16x2 LCD display. If you want the LCD on other lines, that would
have to be specified here.  We are going to use 8-bit mode to transfer
data. All these numbers refer to lines on the Propeller.
}}

  _CLKMODE=XTAL1 + PLL2X           'The system clock spec
  _XINFREQ = 5_000_000             'the oscillator frequency

  RegSelect     = 16
  ReadWrite     = 17
  Enable        = 18
  DataBit0      = 8
  DataBit7      = 15
  waitPeriod    =500_000              'set the wait period in Milliseconds
  high          =1                         'define the High state
  low           =0                         'define the Low state
{{
Defining high and low states will allow us to invert these when we use
buffers to amplify the output from the prop chip.  We will then make
low=1 and high=0 thus inverting all the values throughout the program
}}

PUB Go
  DIRA[DataBit7..DataBit0]:=%11111111  'the lines for the LCD are outputs
  DIRA[RegSelect] := High           'the lines for the LCD are outputs
  DIRA[ReadWrite] := High           'the lines for the LCD are outputs
  DIRA[Enable]    := High           'the lines for the LCD are outputs

  INITIALIZE_LCD                    'initialize the LCD
  waitcnt(5_000_000+cnt)            'wait for LCD to start up
  CLEAR                             'clear the LCD
  repeat                              'repeat forever
    clear
    repeat 4                          'print 4 'A's
      SEND_CHAR ("A")
```

(*continued*)

Program 21-7 Minimal Program to Send Characters to the LCD (*continued*)

```
    repeat 4                              'print 4 'a's
      SEND_CHAR ("b")
    POSITION (1,2)                        'move to POSITION: line 2, space 1
    repeat 4                              'print 4 'B's
      SEND_CHAR ("C")
    repeat 4                              'print 4 'b's
      SEND_CHAR ("d")
    waitcnt(10_000_000+cnt)

PRI INITIALIZE_LCD              'The addresses and data used here are
  waitcnt(500_000+cnt)          'specified in the Hitachi data sheet for
                                'display. YOU MUST CHECK THIS FOR YOURSELF
  OUTA[RegSelect] := Low        'these three lines are specified to write
  OUTA[ReadWrite] := Low               'the initial set up bits for the LCD
  OUTA[Enable]    := Low               'See Hitachi HD44780 data sheet
                                       'YOU MUST CHECK THIS FOR YOURSELF.
  SEND_INSTRUCTION (%0011_0000)        'Send 1st
  waitcnt(49_200+cnt)                  'wait
  SEND_INSTRUCTION (%0011_0000)        'Send 2nd
  waitcnt(1_200+cnt)                   'wait
  SEND_INSTRUCTION (%0011_0000)        'Send 3rd
  waitcnt(12_000+cnt)                  'wait
  SEND_INSTRUCTION (%0011_1000)        'Sets DL=8 bits, N=2 lines, F=5x7 font
  SEND_INSTRUCTION (%0000_1111)        'Display on, Cursor on, Blink on
  SEND_INSTRUCTION (%0000_0001)        'clear LCD
  SEND_INSTRUCTION (%0000_0110)        'Move Cursor, Do not shift display

PUB CLEAR                              'Clear the LCD display and go home
  SEND_INSTRUCTION (%0000_0001)

PUB POSITION (LINE_NUMBER, HOR_POSITION) | CHAR_LOCATION    'Pos crsr
  'HOR_POSITION : Horizontal Position : 1 to 16
  'LINE_NUMBER : Line Number : 1 or 2
  CHAR_LOCATION := (HOR_POSITION-1) * 64  'figure location
  CHAR_LOCATION += (LINE_NUMBER-1) + 128  'figure location
  SEND_INSTRUCTION (CHAR_LOCATION)    'send the instr to position cursor

PUB SEND_CHAR (DISPLAY_DATA)      'set up for writing to the display
  CHECK_BUSY                      'wait for busy bit to clear before sending
  OUTA[ReadWrite] := Low          'Set up to read busy bit
  OUTA[RegSelect] := High         'Set up to read busy bit
  OUTA[Enable]    := High         'Set up to toggle bit H>L
  OUTA[DataBit7..DataBit0] := DISPLAY_DATA    'Ready to SEND data in
  OUTA[Enable]    := Low          'Toggle the bit H>L
```

(*continued*)

Program 21-7 Minimal Program to Send Characters to the LCD (*continued*)

```
PUB CHECK_BUSY | BUSY_BIT      'routine to check busy bit
  OUTA[ReadWrite] := High      'Set to read the busy bit
  OUTA[RegSelect] := Low        'Set to read the busy bit
  DIRA[DataBit7..DataBit0] := %0000_0000
  REPEAT                        'Keep doing it till clear
    OUTA[Enable]  := High       's get ready to toggle H>L this bit
    BUSY_BIT := INA[DataBit7]   'the busy bit is bit 7 of the byte read
    OUTA[Enable]  := Low        'make the enable bit go low for H>L toggle
  WHILE (BUSY_BIT == 1)         'do it as long as the busy bit is 1
  DIRA[DataBit7..DataBit0] := %1111_1111

PUB SEND_INSTRUCTION (DISPLAY_DATA) 'set up for writing instructions
  CHECK_BUSY                   'wait for busy bit to clear before sending
  OUTA[ReadWrite] := Low       'Set up to read busy bit
  OUTA[RegSelect] := Low       'Set up to read busy bit
  OUTA[Enable]    := High      'Set up to toggle bit H>L
  OUTA[DataBit7..DataBit0] := DISPLAY_DATA    'Ready to READ data in
  OUTA[Enable]    := Low        'Toggle the bit H>L
```

Sophisticated Total LCD Control

Now that we have an understanding of how to go about addressing the LCD, we can improve the routines that were created to make them more sophisticated and thus more useful. After we are done with the improvements, we will assign these methods to a separate file from where they can be called by all users and by all the procedures we create. This means that we will not have to include the methods as a part of our other programs. We will call these methods as we need them by including the relevant mother objects under the OBJ declaration in our programs.

The new (8-bit) program, which lists all the new methods, is shown in Program 21-8.

Program 21-8 Comprehensive LCD Control (Demonstration)

```
{{11 Sep 09     Harprit Sandhu
LCDminimal2.spin
Propeller Tool Ver 1.2.6
Chapter 21 Program 8

LCD control.  Comprehensive.

Here are the improvements to program
   Can now send DECIMAL values to the LCD
   Can now send HEX     values to the LCD
   Can now send BINARY  values to the LCD
```

(*continued*)

Program 21-8 Comprehensive LCD Control (Demonstration) (*continued*)

```
   Delays in the print routines have been eliminated to speed things up.
   The output blinks "Hello world" and 1234567890"on the two lines.
}}
CON
  _CLKMODE=XTAL1 + PLL2X           'The system clock spec
  _XINFREQ = 5_000_000             'the oscillator frequency
  DataBit0   =  8          'Data uses 8 bits from
  DataBit7   = 15          'lines 8 to 15
  RegSelect  = 16          'The three control lines, register select
  ReadWrite  = 17          'Read Write and
  Enable     = 18          'Enable  line
  waitPeriod =5_000_000    'set the wait period, about 1/2 sec
  high       =1            'define the High state
  low        =0            'define the Low state

VAR
  long index               'used to count the chars in the string
  long char_index          'used to count the chars in the string

PUB START
   DIRA[16..18]~~                                  'set these 8 lines to outputs
  INITIALIZE_LCD                                   'this initializes LCD
  DIRA[DataBit7..DataBit0] := %1111_1111           '11 lines LCD as outputs
  DIRA[RegSelect] := 1                     'select the register for the LCD
  DIRA[ReadWrite] := 1                               'set to write
  DIRA[Enable]    := 1                               'enable operation
  repeat                                             'this loops forever
    initialize_lcd                                   'init the LCD
    waitCnt(waitPeriod + cnt)                        'wait to see it clear
    position (1,1)                                   ' go to pos 1,1
    PRINT(string("Hello world"))                     'display text message
    POSITION (2, 1)                                  'pos to line 2 position 1
    PRINT_DEC (1234567890)                           'print the number
    waitCnt(waitPeriod + cnt)                        'wait before looping

PRI INITIALIZE_LCD                 'The addresses and data used here are
  waitcnt(500_000+cnt)             'specified in the Hitachi data sheet for the
                                   'display. YOU MUST CHECK THIS FOR YOURSELF.
  OUTA[RegSelect] := 0             'three lines are specified so we can write
  OUTA[ReadWrite] := 0             'the initial set up bits for the LCD
  OUTA[Enable]    := 0             'See Hitachi HD44780 data sheet
  SEND_INSTRUCTION (%0011_1000)    'Sets DL=8 bits, N=2 lines, F=5x7 font
  SEND_INSTRUCTION (%0000_0001)       'clears the LCD
  SEND_INSTRUCTION (%0000_1100)       'Display on, Cursor off, Blink off
```

(*continued*)

Program 21-8 Comprehensive LCD Control (Demonstration) (*continued*)

```
  SEND_INSTRUCTION (%0000_0110)      'Move Cursor, Do not shift display
  {this blank line ends this method}

PRI CLEAR                            'Clear the LCD display and go home
  SEND_INSTRUCTION (%0000_0001)      ' clear screen, go home command
  {this blank line ends this method}

PRI PRINT (the_line)          'routine handles more Chars at a time
                              'called as PRINT(string("the_line")) "
                              'the line" contains the pointer to the line.
                              'because we have to point to the line
                              'zero terminated but we will not use that.
                              'We will use the string size instead.
                              'This was is easier to understand
  index:=0                    'Reset the counter to count chars sent
  repeat                      'repeat for all chars in the list
    char_index:= byte[the_line][index++]   'contains the char/byte
                                           'pointed to by the index
    SEND_CHAR (char_index)                 ' 'pointed to' char to the LCD
  while index<strsize(the_line)            ' till last char is sent

PRI POSITION (LINE_NUMBER, HOR_POSITION) | CHAR_LOCATION    'Pos the crsr
                                                           'at location
  CHAR_LOCATION := (LINE_NUMBER-1) * 64    'fig loc. See Hitachi data sht
  CHAR_LOCATION += (HOR_POSITION-1) + 128 'fig loc. See Hitachi data sht
  SEND_INSTRUCTION (CHAR_LOCATION)         ' position cursor

PRI SEND_CHAR (DISPLAY_CHAR)               'set up for writing to the display
  CHECK_BUSY                               'wait for busy bit to clear
  OUTA[ReadWrite] := 0                           'Set up to read busy bit
  OUTA[RegSelect] := 1                           'Set up to read busy bit
  OUTA[Enable]    := 1                           'Set up to toggle bit H>L
  OUTA[DataBit7..DataBit0] := DISPLAY_CHAR       'Ready to SEND data in
  OUTA[Enable]    := 0                           'Toggle the bit H>L

PRI CHECK_BUSY | BUSY_BIT                        'routine to check busy bit
  OUTA[ReadWrite] := 1                           'Set to read the busy bit
  OUTA[RegSelect] := 0                           'Set to read the busy bit
  DIRA[DataBit7..DataBit0] := %0000_0000         'Set to be an input
  REPEAT                                         'Keep doing it till clear
    OUTA[Enable]  := 1         'set to 1 to toggle H>L this bit
    BUSY_BIT := INA[DataBit7]  'the busy bit is bit 7
                               'INA is the 32 input pins on the PROP, we
                               'are reading data bit 7 on pin 15!
    OUTA[Enable]  := 0         'make the enable  low for H>L toggle
```

(continued)

Program 21-8 Comprehensive LCD Control (Demonstration) (*continued*)

```
  WHILE (BUSY_BIT == 1)                'as long as the busy bit is 1
  DIRA[DataBit7..DataBit0] := %1111_1111  ' data port back to outputs

PRI SEND_INSTRUCTION (DISPLAY_DATA)     'set up for writing instructions
  CHECK_BUSY                            'wait for busy bit to clear
  OUTA[ReadWrite] := 0                          'Set up to read busy bit
  OUTA[RegSelect] := 0                          'Set up to read busy bit
  OUTA[Enable]    := 1                          'Set up to toggle bit H>L
  OUTA[DataBit7..DataBit0] := DISPLAY_DATA      'Ready to READ data in
  OUTA[Enable]    := 0                          'Toggle the bit H>L

PRI PRINT_DEC (VALUE) | TEST_VALUE       'for print vals in deci format
  IF (VALUE < 0)                         'if it is a negative value
    -VALUE                               'change it to a positive
    SEND_CHAR("-")                       'and print a - sign on the LCD
  TEST_VALUE := 1_000_000_000            'we comp to this
                                         'value
  REPEAT 10                              'There are 10 digits maximum
    IF (VALUE => TEST_VALUE)             'see if our num > than testValue
      SEND_CHAR(VALUE / TEST_VALUE + "0")  ' divide to get the digit
      VALUE //= TEST_VALUE                 'figure for the nxt digit
      RESULT~~                             'result so we can pass it on
    ELSEIF (RESULT OR TEST_VALUE == 1)     'then division was even
      SEND_CHAR("0")                       'so we sent out a zero
    TEST_VALUE /= 10                       ' test the next digit

PRI PRINT_HEX (VALUE, DIGITS)        'for printing values in HEX format
  VALUE <<= (8 - DIGITS) << 2        'you can specify up to 8 digits or
                                           'FFFF FFFF max
  REPEAT DIGITS                            'do each digit
    SEND_CHAR(LOOKUPZ((VALUE <-= 4) & $F : "0".."9", "A".."F"))
                                     'use look up table to select character

PRI PRINT_BIN (VALUE, DIGITS)    'for printing values in BINARY format
  VALUE <<= 32 - DIGITS                  '32 digits is the max for our system
  REPEAT DIGITS                                'Repeat for digits desired
    SEND_CHAR((VALUE <-= 1) & 1 + "0")         'send a 1 or a 0
  {this blank line ends this method}
```

Program 21-8 demonstrates that we have learned how to write both text and numbers to the LCD. The routine to write binary numbers (Print Hex) is also included, but it was not demonstrated in the program. Test the program for yourself to confirm that it actually works. The rule for tests is, if you have the time and expertise, test it for yourself. Do not take anyone's word for it.

We will not want to have to include all these routines in every object we write, so let's set up a couple objects in which to store the routines as public methods so they can be called up from any object we write. We will create two objects to do this:

- One to contain all the LCD routines
- One to contain the miscellaneous methods we need from time to time

We will call the two files LCDRoutines and Utilities to make them easy to remember. We can include these in our future objects with the OBJ code by including the lines shown in Program 21-9 in each program.

Program 21-9 Calling Routine with OBJ Block Designations

```
OBJ
 LCD : "LCDRoutines"  'We will be using  METHODS in this program
 UTIL : "Utilities"   'We will be using  METHODS in this program
```

LCDRoutines contains the following methods at this time (we will add more methods as we need them):

- INITIALIZE_LCD
- PRINT (THE_LINE)
- POSITION (LINE_NUMBER, HOR_POSITION) | CHAR_LOCATION
- SEND_CHAR (DISPLAY_CHAR)
- SEND_CHAR (DISPLAY_CHAR)
- PRINT_DEC (VALUE) | TEST_VALUE
- PRINT_HEX (VALUE, DIGITS)
- PRINT_BIN (VALUE, DIGITS)
- CLEAR
- HOME
- SPACE

When we put it all together, we get the code shown in Program 21-10. The routines have been taken from the programs we have developed so far, with a few simple additions. This program is in 8-bit mode.

Program 21-10 Usable LCD Methods in LCDRoutines

```
{{21 Sep 09     Harprit Sandhu
LCDRoutines.spin
Propeller Tool Ver 1.2.6
Chapter 21 Program 10

LCD ROUTINES
```

(continued)

Program 21-10 Usable LCD Methods in LCDRoutines (*continued*)

```
The following are the names of the methods described in this program.
  INITIALIZE_LCD
  INITIALIZE_LCD4 not yet working right.
  PRINT (the_line)
  POSITION (LINE_NUMBER, HOR_POSITION) | CHAR_LOCATION
  SEND_CHAR (DISPLAY_CHAR)
  SEND_CHAR (DISPLAY_CHAR)
  PRINT_DEC (VALUE) | TEST_VALUE
  PRINT_HEX (VALUE, DIGITS)
  PRINT_BIN (VALUE, DIGITS)
  CLEAR
  HOME
  SPACE (QTY)

Revisions
  04 Oct 09  Initialize made more robust, misc unnecessary calls removed.

}}
CON                                'all the constants used by all the METHODS
                                   'in this program have to be listed here
  _CLKMODE=XTAL1+ PLL2X            'The system clock spec
  _XINFREQ = 5_000_000             '10 MHz
  DataBit0    =  8                 'Data uses 8 bits from here on
  DataBit1    =  9
  DataBit2    = 10
  DataBit3    = 11                 'All these bits don't need to be named but
  DataBit4    = 12                 'are named so that they can be called by
  DataBit5    = 13                 'name if the need ever arises
  DataBit6    = 14
  DataBit7    = 15
  RegSelect   = 16                 'The three control lines
  ReadWrite   = 17                 '
  Enable      = 18                 '
  high        =1                   {define the High state}
  low         =0                   {define the Low state}

VAR                                'these are the variables we will use.
   byte  temp                      'for use as a pointer
   byte  index                     'to count characters

PUB Go
  INITIALIZE_LCD
  repeat
    print(String("1234567890  L1"))
    position(2,1)
    print(String("On second line"))
```

(*continued*)

Program 21-10 Usable LCD Methods in LCDRoutines (*continued*)

```
    waitcnt(3_000_000+cnt)
    clear
    waitcnt(3_000_000+cnt)

{{initialize the LCD to use 8 lines of data
Includes a half-second delay, clears the display and positions to 1,1
no variables used
}}
PUB INITIALIZE_LCD                  'The addresses and data used here are
  waitcnt(5_000_000+cnt)            'specified in the Hitachi data sheet
  DIRA[DataBit0..Enable]~~          'YOU MUST CHECK THIS FOR YOURSELF.
  SEND_INSTRUCTION (%0011_0000)     'Send 1
  waitcnt(49_200+cnt)               'wait
  SEND_INSTRUCTION (%0011_0000)     'Send 2
  waitcnt(1_200+cnt)                'wait
  SEND_INSTRUCTION (%0011_0000)     'Send 3
  waitcnt(12_000+cnt)               'wait
  SEND_INSTRUCTION (%0011_1000)     'Sets DL=8 bits, N=2 lines, F=5x7 font
  SEND_INSTRUCTION (%0000_1110)     'Display on, Blink on, Sq Cursor off
  SEND_INSTRUCTION (%0000_0110)     'Move Cursor, Do not shift display
  SEND_INSTRUCTION (%0000_0001)     'clears the LCD
  POSITION (1,1)

{{Sends instructions as opposed to a character to the LCD
no variables are used
}}
PUB SEND_INSTRUCTION (D_DATA)       'set up for writing instructions
  CHECK_BUSY                              'wait for busy bit to clear b
  OUTA[ReadWrite] := 0                    'Set up to read busy bit
  OUTA[RegSelect] := 0                    'Set up to read busy bit
  OUTA[Enable]    := 1                    'Set up to toggle bit H>L
  OUTA[DataBit7..DataBit0] := D_DATA      'Ready to READ data in
  OUTA[Enable]    := 0              'Toggle the bit H>L to Xfer the data

{{Sends a character to the LCD
}}
PUB SEND_CHAR (D_CHAR)              'set up for writing to the display
  CHECK_BUSY                              'wait for busy bit to clear
  OUTA[ReadWrite] := 0                    'Set up to read busy bit
  OUTA[RegSelect] := 1                    'Set up to read busy bit
  OUTA[Enable]    := 1                    'Set up to toggle bit H>L
  OUTA[DataBit7..DataBit0] := D_CHAR       'Ready to SEND data in
  OUTA[Enable]    := 0                    'Toggle the bit H>L

{{Print a line of characters to the LCD
uses variables index and temp
}}
```

(continued)

Program 21-10 Usable LCD Methods in LCDRoutines (*continued*)

```
PUB PRINT  (the_line)               'This routine handles more than one Char
                                    'called as PRINT(string("the_line"))
                                    '"the_line" contains pointr to line
                                    'because we have to point to the line
                                    'zero terminated but will not use that.
                                    'use the string size instead. Easier
  index:=0                          'Reset the counter we are using
  repeat                            'repeat for all chars in the list
    temp:= byte[the_line][index++]'temp contains the char pointed by index
    SEND_CHAR (temp)                'send the 'pointed to' char to the LCD
  while index<strsize(the_line)     ' till the last char is sent

{{Position cursor
}}
PUB POSITION (LINE_NUMBER, HOR_POSITION) | CHAR_LOCATION  'Pos the crsr
  'Horizontal Position : 1 to 16         'specified by the two numbers
  'Line Number : 1 or 2
  CHAR_LOCATION := (LINE_NUMBER-1) * 64   'figr loc. See Hitachi  data
  CHAR_LOCATION += (HOR_POSITION-1) + 128 'figr loc. See Hitachi  data
  SEND_INSTRUCTION (CHAR_LOCATION)       'send the instr to pos crsr

{{Check for busy
}}
PUB CHECK_BUSY | BUSY_BIT            'routine to check busy bit
  OUTA[ReadWrite] := 1               'Set to read the busy bit
  OUTA[RegSelect] := 0               'Set to read the busy bit
  DIRA[DataBit7..DataBit0] := %0000_0000 'Set to be an input
  REPEAT                             'Keep doing it till clear
    OUTA[Enable]  := 1               'set to 1 to toggle H>L this bit
    BUSY_BIT := INA[DataBit7]        'the busy bit is bit 7 of the byte read
                        'INA is the 32 input pins on the PROP and we
                        'are reading data bit 7 which is on pin 15!
    OUTA[Enable]  := 0  'make the enable bit go low for H>L toggle
  WHILE (BUSY_BIT == 1) 'do it as long as the busy bit is 1
  DIRA[DataBit7..DataBit0] := %1111_1111  'set back to outputs

{{Print decimal
}}
PUB PRINT_DEC (VALUE) | TEST_VALUE'for printing values in decimal format
  IF (VALUE < 0)                    'if it is a negative value
    -VALUE                          'change it to a positive
    SEND_CHAR("-")                  'and print a - sign on the LCD
                                    '
  TEST_VALUE := 1_000_000_000  'we get indiv digits by comp to this
                               'value. div by 10 to get the nxt val
  REPEAT 10                    'T 10 digits maximum in our system
```

(*continued*)

Program 21-10 Usable LCD Methods in LCDRoutines (*continued*)

```
    IF (VALUE => TEST_VALUE)    'see if bigger than testValue
      SEND_CHAR(VALUE / TEST_VALUE + "0")      'divide to get the digit
      VALUE //= TEST_VALUE              'figure for the next digit
      RESULT~~                          'result we can pass it on below
    ELSEIF (RESULT OR TEST_VALUE == 1) 'if then div was even
      SEND_CHAR("0")                   'so we sent out a zero
    TEST_VALUE /= 10        'we divide by 10 to test for the next digit

{{Print Hex
}}
PUB PRINT_HEX (VALUE, DIGITS)      'for printing values in HEX format
  VALUE <<= (8 - DIGITS) << 2      ' specify up to 8 digits, FFFFFFFF max
  REPEAT DIGITS                    'do each digit
    SEND_CHAR(LOOKUPZ((VALUE <-= 4) & $F : "0".."9", "A".."F"))
                                   'use lookup table to select character

{{Print Binary
}}
PUB PRINT_BIN (VALUE, DIGITS)  'for printing values in BINARY format
  VALUE <<= 32 - DIGITS   '32 binary digits is the max for our system
  REPEAT DIGITS                    'Repeat for each digit desired
    SEND_CHAR((VALUE <-= 1) & 1 + "0")       'send a 1 or a 0

{{Clear screen
}}
PUB CLEAR                         'Clear the LCD display and go home
  SEND_INSTRUCTION (%0000_0001)  ' clear screen and go home command

{{Go to position 1,1
}}
PUB HOME                           'go to position 1,1.
  SEND_INSTRUCTION (%0000_0011)    'Not cleared

{{Print spaces
}}
PUB SPACE (qty)                    'Prints spaces, for between numbers
  repeat (qty)
    PRINT(STRING(" "))
```

In order to make the LCD routines useful, they need to be amenable to thorough testing from time to time. To do that, we need a program that tests all the routines automatically and then automatically displays what is going on in the LCD (see Program 21-11). This will be especially useful when we have to make a change to a routine and need to ensure that it still works with everything else the way we intended.

Program 21-11 Preliminary LCD Routines-Checking Program

```
{{11 Oct 09     Harprit Sandhu
LCDTester.Spin
Propeller Tool Ver. 1.2.6

Test program for LCD Routines
}}

CON
  _CLKMODE=XTAL1+ PLL2X          'The system clock spec
  _XINFREQ = 5_000_000
  high       =1             'define the High state
  low        =0             'define the Low state

VAR
  byte index                'used as counter for various uses

OBJ
  LCD : "LCDRoutines"       'Using their METHODS in this program
  UTIL : "Utilities"        'Using their METHODS in this program

PUB Go

LCD.initialize_lcd
LCD.print (string("Initialized OK"))
LCD.position(2,1)
LCD.print (string("Print test"))
UTIL.pause (1500)

LCD.clear
LCD.position (1,1)
LCD.print (string("this is 1st line"))
UTIL.pause (300)
LCD.position(2,1)
LCD.print (string("ok is on line 2"))
UTIL.pause (1500)

LCD.clear
LCD.print (string("Position test"))
UTIL.pause (1500)
index:=0
repeat 16
  index := index +1
  LCD.clear
  LCD.position (1,index)
  LCD.print (string("X"))
  UTIL.pause (300)
```

(continued)

Program 21-11 Preliminary LCD Routines-Checking Program (*continued*)

```
index:=0
repeat 16
  index := index +1
  LCD.clear
  LCD.position (2,index)
  LCD.print (string("X"))
  UTIL.pause (300)
LCD.clear
LCD.print (string("End"))
Repeat
```

As you become more proficient in the use of Spin, you will want to make Program 21-11 more comprehensive and robust to check all the features you add to the LCDRoutines program. Program 21-11 is just the beginning.

The Utilities program will contain the following two methods for now (we will add more methods as we need them):

- **FLASH** Pulses a line (usually an LED)
- **PAUSE** Pause/delay program in milliseconds

The code for the Utilities program is given in Program 21-12.

Program 21-12 Utilities

```
{{21 Sep 09    Harprit Sandhu
Utilities0.spin
Propeller Tool Ver 1.2.6
Chapter 21 Program 12

Program UTILITIES
  Flash         flashes a pin once, toggles it slowly
  Pause         pause in milliseconds

  }}
CON
  high          =1
  low           =0
  waitperiod    =100

PUB FLASH (color)                         'routine to flash an LED by color
    outa[color] :=high                    'line that actually sets the LED high
    waitCnt(waitPeriod +cnt)              'wait till counter reaches this value
    outa[color] :=low                     'line that actually sets the LED low
    waitCnt(waitPeriod +cnt)              'wait till counter reaches this value
                                          '
PUB PAUSE(millisecs)        'As set up here it is .25 millisceconds
  waitcnt((clkfreq/1000)*millisecs +cnt)      'based on Osc freq
```

This completes what we need to have usable control of a 16-character-by-2-line LCD display in 8-bit mode. We will be using this 16×2 display for all our experiments, but we will use 4-bit mode so that we can free up four more lines on the Propeller. Four-bit mode is covered next.

4-Bit Mode

The use of the LCD in 4-bit mode has a few complications we need to understand. Because most data is expressed in 8 bits and only four data lines are connected to the LCD, some special procedures have to be followed to allow all 8 bits to be transferred to the LCD. In general, this is done by transferring the data in two steps, with 4 bits transferred in each step.

When the LCD powers up, its condition is indeterminate. The first few instructions can be sent to it on four lines or on eight lines, but only bits 7 to 4 will be looked at by the LCD. The first three transmissions can be sent as %0011_0000 or as %0011.

The fourth transmission has to be %0010 to set 4-bit mode. This is the last 4-bit transmission, because at the end of this transmission, the system is in 4-bit mode. In 8-bit mode, this would have been the high part of the %0011_1000 instruction (in 4-bit mode, this is %0010_1000—the 0 that replaces the 1 is the 4-bit mode specification). This means that the low nibble (%1000) has been ignored in the process. We pick this byte back up by transmitting the entire 8 bits (%0010_1000) again 4 bits at a time.

The SEND_INSTRUCTION used in 8-bit mode is divided into two instructions in the LCDRoutines4 object:

- **SEND_INSTRUCTION** sends just 4 bits.
- **SEND_INSTRUCTION2** sends all 8 bits, 4 bits at a time.

All the routines in LCDRoutines that transmit 8 bits of data are modified to send 4 bits at a time for both characters and data. All these changes are incorporated into the LCDRoutines4 program listed in Program 21-13.

Note *Having accomplished this change to the LCD, in the future we will use 4-bit mode in all our experiments.*

Program 21-13 Listing for Program LCDRoutines4

```
{{21 Sep 09      Harprit Sandhu
LCDRoutines4.spin
Propeller Tool Version 1.2.6
Chapter 21 Program 13 and Appendix

LCD ROUTINES for a 4-bit data path.
```

(*continued*)

Program 21-13 Listing for Program LCDRoutines4 (*continued*)

```
The following are the names of the methods described in this program.

  INITIALIZE_LCD
  PRINT (the_line)
  POSITION (LINE_NUMBER, HOR_POSITION) | CHAR_LOCATION
  SEND_CHAR (DISPLAY_CHAR)
  SEND_CHAR (DISPLAY_CHAR)
  PRINT_DEC (VALUE) | TEST_VALUE
  PRINT_HEX (VALUE, DIGITS)
  PRINT_BIN (VALUE, DIGITS)
  CLEAR
  HOME
  SPACE (QTY)

Revisions
  04 Oct 09  Initialize made more robust, misc unnecessary calls removed.

}}
CON                                 'all the constants used by all the METHODS
                                    'in this program have to be listed here
  _CLKMODE=XTAL1 + PLL2X            'The system clock spec. 2 X multiplier
  _XINFREQ    = 5_000_000           'ext crystal is 5 MHz, so 10 MHz operation
  DataBit4    = 12                  'are named so that they can be called by
  DataBit5    = 13                  'name if the need ever arises
  DataBit6    = 14                  '
  DataBit7    = 15                  '
  RegSelect   = 16                  'The three control lines
  ReadWrite   = 17                  'The three control lines
  Enable      = 18                  'The three control lines
  high        =1                    'define the High state
  low         =0                    'define the Low state
  Inv_high    =0                    'define the Inverted High state
  Inv_low     =1                    'define the Inverted Low state

VAR                                 'these are the variables we will use
   byte  temp                       'for use as a pointer
   byte  index                      'to count characters

PUB Go
  INITIALIZE_LCD
  repeat
    print(String("4bit mode line 1"))
    position(2,1)
    print(String("4bit mode line 2"))
    waitcnt(clkfreq/4+cnt)
    clear
```

(continued)

Program 21-13 Listing for Program LCDRoutines4 (*continued*)

```
    waitcnt(clkfreq/4+cnt)

{{initialize the LCD to use 4 lines of data
Includes the half-second delay, clears the display and positions to 1,1
no variables used
}}
PUB INITIALIZE_LCD              'The addresses and data used here are
  waitcnt(150_000+cnt)          'specified in the Hitachi data sheet for the
  DIRA[DataBit4..Enable]~~      'display. YOU MUST CHECK THIS FOR YOURSELF.
  SEND_INSTRUCTION (%0011)      'Send 1st
  waitcnt(49_200+cnt)           'wait
  SEND_INSTRUCTION (%0011)      'Send 2nd
  waitcnt(1_200+cnt)            'wait
  SEND_INSTRUCTION (%0011)      'Send 3rd
  waitcnt(12_000+cnt)           'wait
  SEND_INSTRUCTION (%0010)      'set for 4 bit mode
  waitcnt(12_000+cnt)           'wait

  SEND_INSTRUCTION2 (%0010_1000)  'Sets DL=4 bits, N=2 lines, F=5x7 font
 ' waitcnt(12_000+cnt)            'wait
  SEND_INSTRUCTION2 (%0000_1100)  'Display on, Blink on, Sq Cursor off
 ' waitcnt(12_000+cnt)            'wait
  SEND_INSTRUCTION2 (%0000_0110)  'Move Cursor, Do not shift display
 ' waitcnt(12_000+cnt)            'wait
  SEND_INSTRUCTION2 (%0000_0001)  'clears the LCD
  'waitcnt(12_000+cnt)            'wait
  POSITION (1,1)

{{Sends instructions as opposed to a character to the LCD
no variables are used
}}
PUB SEND_INSTRUCTION (D_DATA)   'set up for writing instructions
  CHECK_BUSY                    'wait for busy bit to clear before sending
  OUTA[ReadWrite] := 0          'Set up to read busy bit
  OUTA[RegSelect] := 0          'Set up to read busy bit
  OUTA[Enable]    := 1          'Set up to toggle bit H>L
  OUTA[DataBit7..DataBit4] := D_DATA 'Ready to READ data in
  OUTA[Enable]    := 0                'Toggle the bit H>L to Xfer the data

{{Sends an instruction as opposed to a character to the LCD
no variables are used
}}
PUB SEND_INSTRUCTION2 (D_DATA)  'set up for writing instructions
  CHECK_BUSY          'wait for busy bit to clear before sending
  OUTA[ReadWrite] := 0          'Set up to read busy bit
```

(*continued*)

Program 21-13 Listing for Program LCDRoutines4 (*continued*)

```
  OUTA[RegSelect] := 0          'Set up to read busy bit
  OUTA[Enable]    := 1          'Set up to toggle bit H>L
  OUTA[DataBit7..DataBit4] := D_DATA>>4   'Ready to READ data in
  OUTA[Enable]    := 0     'Toggle the bit H>L to Xfer the data
  OUTA[Enable]    := 1                  'Set up to toggle bit H>L
  OUTA[DataBit7..DataBit4] := D_DATA   'Ready to READ data in
  OUTA[Enable]    := 0     'Toggle the bit H>L to Xfer the data

{{Sends a single character to the LCD in two halves
}}
PUB SEND_CHAR (D_CHAR)             'set up for writing to the display
  CHECK_BUSY         'wait for busy bit to clear before sending
  OUTA[ReadWrite] := 0          'Set up to send data
  OUTA[RegSelect] := 1          'Set up to send data
  OUTA[Enable]    := 1          'go high
  OUTA[DataBit7..DataBit4] := D_CHAR>>4  'Send high 4 bits
  OUTA[Enable]    := 0                  'Toggle the bit H>L
  OUTA[Enable]    := 1                  'go high again
  OUTA[DataBit7..DataBit4] :=D_CHAR     'send low 4 bits
  OUTA[Enable]    := 0                  'Toggle the bit H>L

{{Print a line of characters to the LCD
uses variables index and temp
}}
PUB PRINT (the_line)   'This routine handles more than one Char at a time
                       'called as PRINT(string("the_line"))
                       '"the_line" contains the pointer to line. Line is
                       'because we have to point to the line
                       'zero terminated but we will not use that.  We will
                       'use the string size instead. Easier to understand
  index:=0             'Reset the counter we are using to count chars sent
  repeat               'repeat for all chars in the list
    temp:= byte[the_line][index++] 'temp contns char pointed by index
    SEND_CHAR (temp)               'send the 'pointed to' char to the LCD
  while index<strsize(the_line)    ' till the last char is sent

{{Position cursor
}}
PUB POSITION (LINE_NUMBER, HOR_POSITION) | CHAR_LOCATION  'Line Number : 1 to 4
  'Horizontal Position : 1 to 20              'specified by the two numbers
  CHAR_LOCATION := (LINE_NUMBER-1) * 64     'figure location. See Hitachi
  CHAR_LOCATION += (HOR_POSITION-1) + 128 'figure location. See Hitachi
 ' CHAR_LOCATION += (HOR_POSITION-1) + 128
 ' CHAR_LOCATION += (HOR_POSITION-1) + 128
  SEND_INSTRUCTION2 (CHAR_LOCATION)        ' instr to position cursor
```

(continued)

Program 21-13 Listing for Program LCDRoutines4 (*continued*)

```
{{Check for busy
}}
PUB CHECK_BUSY | BUSY_BIT                    'routine to check busy bit
  OUTA[ReadWrite] := 1                       'Set to read the busy bit
  OUTA[RegSelect] := 0                       'Set to read the busy bit
  DIRA[DataBit7..DataBit4] := %0000          'Set the entire port to be an input
  REPEAT                                  'Keep doing it till clear
    OUTA[Enable]  := 1  'set to 1 to get ready to toggle H>L this bit
    BUSY_BIT := INA[DataBit7]  'the busy bit is bit 7 of the byte read
                            'INA is the 32 input pins on the PROP and we
                            'are reading data bit 7 which is on pin 15!
    OUTA[Enable]  := 0      'make the enable bit go low for H>L toggle
  WHILE (BUSY_BIT == 1)     'do it as long as the busy bit is 1
  DIRA[DataBit7..DataBit4] :=%1111 'done, so set the data port back to outputs

{{Print a decimal value, whole numbers only
}}
PUB PRINT_DEC (VALUE) | TEST_VALUE   'for decimal format
  IF (VALUE < 0)                   'if it is a negative value
    -VALUE                         'change it to a positive
    SEND_CHAR("-")                 'and print a - sign on the LCD
  TEST_VALUE := 1_000_000_000  'we get individual digits by compar to this
                             'value and then dividing by 10 to get next val
  REPEAT 10                  'There are 10 digits maximum in system
    IF (VALUE => TEST_VALUE) 'see if our number is > than testValue
      SEND_CHAR(VALUE / TEST_VALUE + "0")   'divide to get the digit
      VALUE //= TEST_VALUE   'figure the next value for the next digit
      RESULT~~               'result of what just did pass it on below
    ELSEIF (RESULT OR TEST_VALUE == 1) ' a 1 then div was even
      SEND_CHAR("0")                'so we sent out a zero
    TEST_VALUE /= 10                ' test for the next digit

{{Print a Hexadecimal value
}}
PUB PRINT_HEX (VALUE, DIGITS)      'for printing values in HEX format
  VALUE <<= (8 - DIGITS) << 2      ' up to 8 digits or FFFFFFFF max
  REPEAT DIGITS                    'do each digit
    SEND_CHAR(LOOKUPZ((VALUE <-= 4) & $F : "0".."9", "A".."F"))
                                   'use lookup table to select character

{{Print a Binary value
}}
PUB PRINT_BIN (VALUE, DIGITS)  'for printing values in BINARY format
  VALUE <<= 32 - DIGITS        '32 binary digits is the max for our sys
  REPEAT DIGITS                         'Repeat for each digit desired
    SEND_CHAR((VALUE <-= 1) & 1 + "0") 'send a 1 or a 0
```

(*continued*)

Program 21-13 Listing for Program LCDRoutines4 (*continued*)

```
{{Clear screen
}}
PUB CLEAR                              'Clear the LCD display and go home
  SEND_INSTRUCTION2 (%0000_0001)       ' clear screen and go home

{{Go to position 1,1   Does not clear the screen
}}
PUB HOME                                   'go to position 1,1.
  SEND_INSTRUCTION2 (%0000_0011)           'LCD not cleared

{{Print spaces
}}
PUB SPACE (qty)                            'Prints spaces, for between numbers
  repeat (qty)
    PRINT(STRING(" "))
```

These routines have to be maintained and improved upon as time goes by. They will reflect how you use the Propeller and what code you borrow from the work of others. This will continue to be a work in progress.

22

RUNNING MOTORS: A PRELIMINARY DISCUSSION

There are times when you need to move something as a part of what you need to do with a microcontroller. The easiest way to do this, of course, is with a small motor or sometimes a solenoid. In this chapter we discuss the scope of what we will be covering in the upcoming chapters.

Note *Running large motors is very much like running small motors, except you need larger amplifiers and you have to provide for more safety interlocks because a lot of energy is being handled. Therefore, you have to do everything possible to keep things from getting out of hand.*

The control of the following types of motors will be covered:

- R/C hobby servos
- Stepper motors (bipolar)
- Small Permanent Magnet Direct Current motors
- DC motors with encoders attached for feedback
- Relays, solenoids, and AC motors of about a quarter horsepower

The control of each of these items is covered in a specific chapter devoted to that topic.

R/C Hobby Servomotors

Large servos weighing a few pounds or more are available to run off larger power supplies and the signals received from hobby radio transmitters. The techniques for running these giant servos are the same as those for running hobby servos used by the hobby radio control industry. As always, safety becomes a serious consideration when we migrate to larger devices needing higher currents and voltages.

Of all the motors we can control with a microcontroller, the easiest to control are the servomotors used by the radio control hobbyists. These motors have integral gearboxes built into them to allow a movement of about 180 degrees at the output shaft. An internal potentiometer allows the system to determine the position of the output shaft. Control is affected by sending the motor a signal pulse of a required duration of about 1.5 microseconds about 60 times a second.

An R/C servo can be positioned to one part in 256 if an 8-bit variable is used to control the width of the pulse sent to it, or one part in 4,096 if a 12-bit variable is used for control. For most purposes served by these small servos, the 8-bit signal is more than adequate for the job.

The control of R/C servos is covered in Chapter 24.

Stepper Motors (Bipolar)

Stepper motors move a part of a revolution with each control signal change. They typically contain an arrangement of magnets and coil windings that allow us to move the motor incrementally (in steps). Motors with 400 steps per revolution or 0.9 degrees per step can be obtained at a reasonable cost. Motors with 200 and fewer steps per revolution are more common and usually cheaper. If the motor is not allowed to be overloaded (and thus to slip), we can keep track of the position of the motor by keeping track of how many times we have sent it a control signal change.

We will cover the control of stepper motors that have four wires, meaning that they have two sets of windings, in detail. These are called bipolar motors, and we will use one that needs 12 volts at about 1 ampere for our experiments. The one I used needed 100 steps to make one complete revolution. The Xavien amplifier we use for our other motors will be able to control this motor, so no other expense is involved. Any bipolar motor can be used that has voltages and amperage characteristics that can be provided by the amplifier we are using (up to 3 amps at between 12 and 55 volts).

Other types of stepper motors are controlled with similar schemes but require more complicated power supplies and electronics. The four-wire motors require only one power supply, and the amplifiers we will use for the DC motors can run them. Each amplifier has to be able to control two sets of motor windings, and most "H" bridge-type dual amplifiers have this capability.

The control of stepper motors is covered in Chapter 26.

Small Brush-Type DC Motors

The control of small or even tiny DC motors can vary from on/off control to simple speed control based on a pulse width modulated (PWM) power signal. The techniques for doing this are covered in Chapter 25. As we try to control larger (but still small) motors, the power needed increases and so does the need for larger electronic controllers to meet the larger power needs. Integrated circuits that allow us to manage the control of these motors are now available from a number of manufacturers. We will cover the use of one of these integrated circuits (the LMD 18200) to control a motor in detail. A number of suppliers provide ready-made amplifiers that use these integrated circuits for controlling small motors. We will discuss a number of the amplifiers available and will investigate how we can use them to control our motors with a microcontroller. The size of the motor we control is limited by the size of the amplifier available to control it. We will limit ourselves to 3 amperes at between 12 volts and 55 volts DC, as provided by the LMD 18200. Because integrated circuits that can handle these amperages and voltages are available at a reasonable cost, we will use an amplifier with these characteristics for all our small motors. Three inexpensive and readily available amplifiers are discussed in Chapter 23.

We will not consider brushless motors. These are similar to brushed motors but use solid-state electronic switching to emulate the function provided by the commutator in a standard DC motor with brushes. In most cases, their control is similar to the control of the usual brushed motor.

DC Motors with Attached Encoders

In order to control both speed and distance traveled (in revolutions), you need some sort of feedback mechanism that will tell you how fast the motor is moving and how far the motor has moved. This is usually done with an optical encoder that provides a quadrature signal of a fixed number of cycles per revolution. We count how many cycles have gone by to determine how far the motors has moved and adjust the power fed to the motor as needed for the results we are trying to achieve. The encoder signal also has the ability to be interpreted for motor direction. The ability of the Propeller to use one cog to keep track of the encoder counts makes controlling servomotors considerably easier in a multiprocessor Propeller-based control scheme.

DC servomotors are covered in Chapter 28.

Relays and Solenoids

Although not strictly motors, relays and solenoids use the same magnetic technology that we are using to control the motors to make small movements. These movements are often useful for the experimenter, and we will cover the techniques needed to control these devices without damaging the sensitive electronics in our microcontroller in Chapter 29.

Small A/C Motors at 120 Volts, Single Phase

Oftentimes, it is necessary to control a small AC motor as a part of what we are doing. Controlling the on/off operation of a small (under a quarter horsepower) AC motor is quite straightforward and will be covered in Chapter 29.

Controlling the speed of an AC motor is a little more complicated because these motors are not designed for speed control. They run at the speed determined by the frequency of the power lines, so the easiest way to vary the speed is by varying the frequency of the power sent to the motor. Even so, there is a limit to how much the speed can be changed because of overheating in the motor windings and resonances related to the overall design of the motor. Controls needed for these types of motors are beyond the scope of a book for beginners.

Understanding the Concept of the "Response Characteristics" of a Motor

An important concept we need to understand is the motor's ability to comply with the commands we send it. This is known as its "response characteristics." If the motor cannot possibly do what we are commanding it to do, we are essentially wasting time because trying to control a motor under such circumstances is meaningless. No matter what we command the motor to do, if the load on the motor, the characteristics of the motor, its power supply, or the controller does not allow the motor to do what we want it to do, no amount of expertise on our part is going to make a difference. Although this seems to be obvious, it is the reason that most design failures occur.

So What Does "Compliance" Mean?

In a nutshell, "compliance" means that the motor has the power to execute the commands sent to it by the controller in real time. Keeping in mind that everything takes some time, *real time* means "right after the motor gets the command" or "immediately." Usually this becomes a problem when the motor is too weak, the load is too large, or the power supply is inadequate for the task being commanded. (The processor we are using has to be up to the task, too. Its software has to have the right commands, and it has to be fast enough to do the job.)

Compliance also means that you have to select a control situation that can be realized if you are going to be successful in your control attempt. It does not, however, mean that we understand the difference between what can and cannot be done at this stage of our learning process. Hopefully, by the time we get to the end of the exercises, we will have a better understanding of what is possible and what is not.

DC Motor Operation Notes

As a general rule, a DC motor in a servo control situation needs to be running (under load) at well below 50% of the power needed to perform the task at hand. The other 50% or so of the power is reserved for the power needed for sudden load changes, to accelerate, and to transition between moves quickly. There will be cases when even more power is needed for short periods. Keep in mind that a DC motor can provide a lot more than its rated power as the supply voltage is increased. Such overloads should be limited to short periods to prevent internal overheating.

The limiting condition for a DC motor is the heat that builds up in the motor windings and the amperage the brushes can transmit to the commutator. If the motor can be kept cool or if the motor materials will handle higher temperatures. DC motors can be pushed well beyond published ratings, especially for short periods of time. We have to monitor the temperature of the windings to make sure things don't get out of hand. Keep in mind that a little overheating over a long period of time can be just as damaging as a lot of heating in a short amount of time. The capacity of the commutator-brush current is a function of its mechanical design. If it is being exceeded, the commutator starts to pit and then the brushes fail.

The top speed of the motor is limited by the back Electro Motive Force that it generates as its speed increases. When the back EMF gets high enough, the motor can no longer increase its speed. At the higher amperages that these speeds require, the ability of the brushes to transmit the necessary current to the commutator is compromised and sparking at the brushes increases. This sparking ruins the commutator and the brushes rather rapidly.

For our purposes, we can consider the motor's response to the voltage applied to be linear. This means that, in general, twice the voltage will give us twice the speed. We will assume that we are working within electrical parameters that the motors can tolerate without difficulty.

23

MOTOR AMPLIFIERS FOR SMALL MOTORS

Note *The output from a Transistor-Transistor Logic gate with a "fan out" of 10 or so will run the smallest of DC motors without amplification. All other motors need at least a small transistor to provide enough energy to run them. For serious motor control, we need an integrated system that provides all the function of an H bridge so that we can implement pulse width modulated motor speed control and motor reversal. A number of integrated circuits that do this are available from a host of vendors, both as integrated circuits and as complete small amplifiers.*

Let's take a look at three small amplifiers that we can use to control our motors. All three are inexpensive and easy to use. Other suppliers provide similar amplifiers you might find more suitable for your particular use, but this tutorial does not cover the use of any other amplifiers. However, you should not have any trouble using these similar amplifiers. Although I describe two amplifiers, I provide the circuitry for only the larger, more powerful Xavien dual amplifier.

The three amplifiers shown in Figure 23-1 each provide an inexpensive way to run one or two motors, coils, or other devices at the same time. Because we need a two-axis amplifier to run the stepper motor, and if you are going to buy only one amplifier, buy one of these two amplifiers. The Solarbotics amplifier is cheaper but also handles fewer amps and you do have to assemble it (see Table 23-1). Most users will find that their amplifier of choice is the *Xavien two axis amplifier,* shown on the right in Figure 23-1.

All these amplifiers accept CMOS or TTL level signals to control the power to the motors. Each requires a power supply that matches the power needed by the motor and the needs of the amplifier electronics. The power supply of the microcontroller and the power supply of the motor/amplifier should be kept separate under all circumstances, with only a common ground connection. If this is not done, noise from the motors will contaminate the power to the microprocessor and cause problems. Motors are very noisy as far as computer electronics are concerned and must be isolated. Motor noise comes from the motor commutators and from the rapid on and off switching of the motor coils.

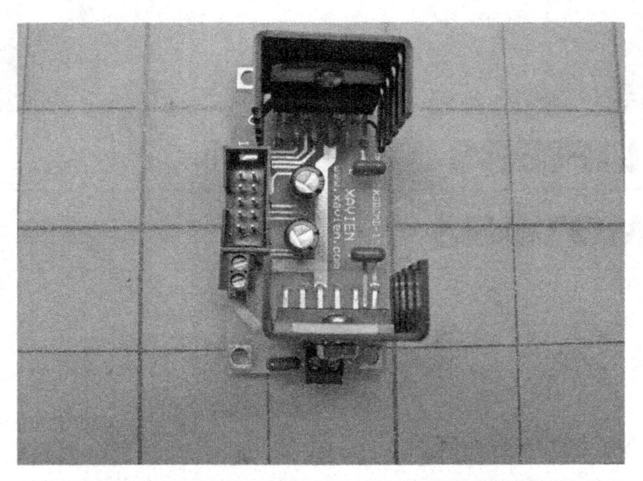

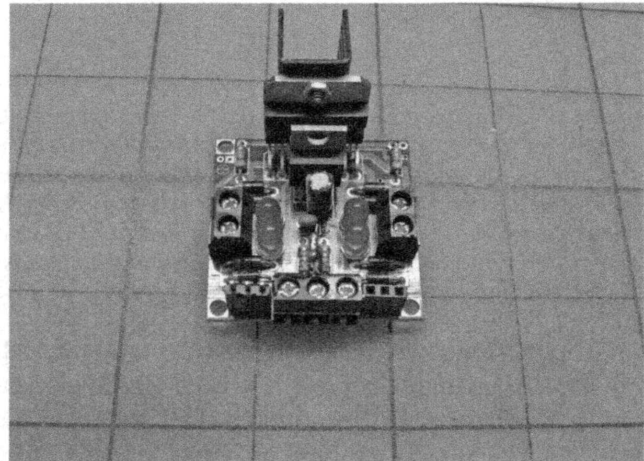

Figure 23-1 Small amplifiers suitable for running small motors

TABLE 23-1 TABLE OF AMPLIFIER PROPERTIES

AMPLIFIER	CHIPS USED	MAX. VOLTAGE	AMPERAGE	COMMENTS
Solarbotics two-axis	L298	6–46	2 amps each axis	Inexpensive kit
Xavien two-axis	LMD18200	12–55	3 amps each axis	Amplifier of choice
Xavien one-axis	MC33886DH	5–28	3.5 amps	Good for low voltages

Although the addition of capacitors to ground from each motor terminal and across the terminals does help, this does not work as well as a well-isolated layout. Because we have a choice, we will use separate power supplies for all our experiments.

Each of these amplifiers uses one or two integrated circuits as its amplifier elements, and the smaller one provides ancillary LEDs and other circuitry to allow interfacing to the signals that the microcontroller provides without an intermediate device. We can go directly from a microprocessor port to the amplifier input without any intermediate electronics.

The capacities of the amplifiers are detailed in Table 23-1.

Amplifier Construction Notes (for Homemade Amplifiers)

Although you can make your own homebrew H bridge amplifier, I do not recommend that you do this—other than as an interesting exercise. The amplifiers you are likely to make (a number of designs are available on the Internet) are likely to be fairly straightforward "H" bridges with no refinements. Unless considerable more sophisticated circuitry is added to the basic amplifier circuit, it is very easy to blow up an H bridge by turning on both transistors on any one side of the bridge at the same time. Trust me: Homebrew amplifiers are unbelievably easy to destroy.

On the other hand, if we use purchased integrated circuits to build our amps, they will almost certainly have circuitry within them to prevent short circuiting, to shut down on overheating, and other useful features. The more sophisticated circuits also provide the ability to detect thermal shutdown and look at the current flow through each amplifier.

Because a number of vendors are willing to sell us ready-to-use amplifiers, at very reasonable prices, there is no good reason (at this stage in our learning process) to not use these resources to run our motors. All discussions and circuits in the tutorial reflect this.

Tip *Reduce the stress in your life. Buy an amplifier.*

Detailed "Use Information" for the Xavien Two-Axis Amplifier

Figure 23-2 shows the Xavien two-axis amplifier. Each of the two 18200 amp chips on this unit can handle up to 3 amps at 55 VDC. Short pulses of 6 amps will be tolerated.

Figure 23-3 shows the connection schematic for the Xavien two-axis amplifier. The polarity of the power to the amplifier is critical and *must be observed.* No other protection is provided. A common ground to the Propeller is required.

Table 23-2 shows the Xavien amplifier control lines.

We will not be using the Current Sense and Thermal Flag lines on lines 4, 5, 9, and 10 in our experiments. However, we *cannot ignore the Brake lines* because they have to be pulled down to turn the brake off.

Figure 23-2 Photograph of the Xavien two-axis amplifier

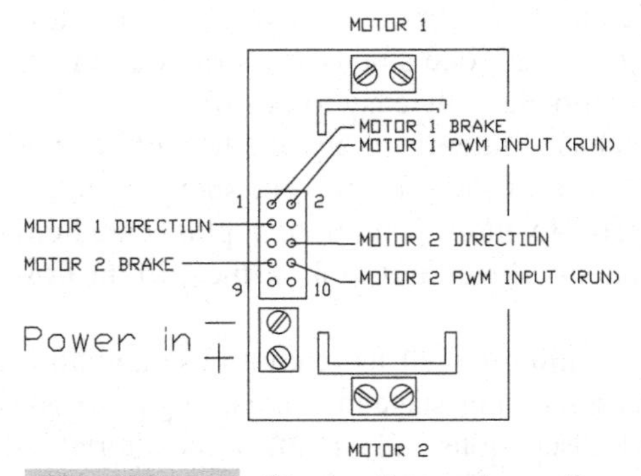

Figure 23-3 Connection schematic for Xavien two-axis amplifier

TABLE 23-2 XAVIEN TWO-AXIS AMPLIFIER OPERATION TABLE

PIN	FUNCTION
1	Motor 1 Brake.
2	Motor 1 PWM or Enable/Run.
3	Motor 1 Direction.
4	Motor 1 Current Sense. (Not used in the discussions.)
5	Motor 1 Thermal Flag. (Not used in the discussions.)
6	Motor 2 Direction.
7	Motor 2 Brake.
8	Motor 2 PWM or Enable/Run.
9	Motor 2 Thermal Flag. (Not used in the discussions.)
10	Motor 2 Current Sense. (Not used in the discussions.)

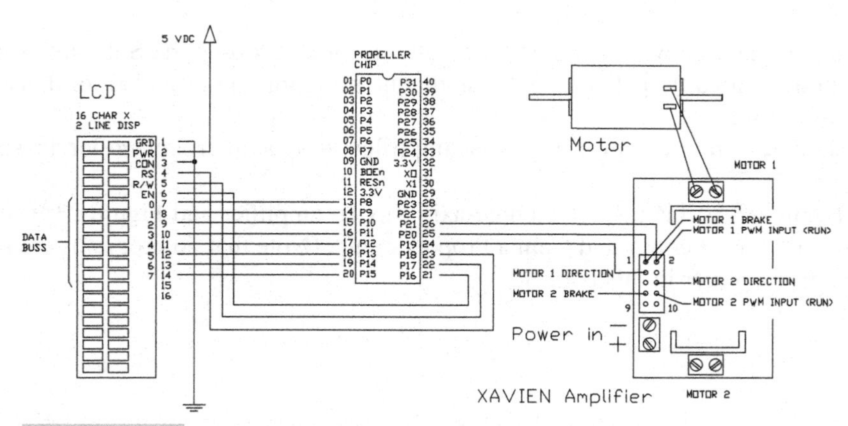

Figure 23-4 **Using the Xavien two-axis amplifier with a motor**

Therefore, pins 1, 2, and 3 can be used to control a coil or motor 1, and pins 6, 7, and 8 will control a coil or motor 2. Figure 23-4 shows one possible scheme for connecting a motor to an amplifier (note that this is not used in our experiments).

Detailed "Use Information" for the Solarbotics Two-Axis Amplifier

Figure 23-5 shows the two-axis Solarbotics amplifier, and Figure 23-6 shows its wiring connections.

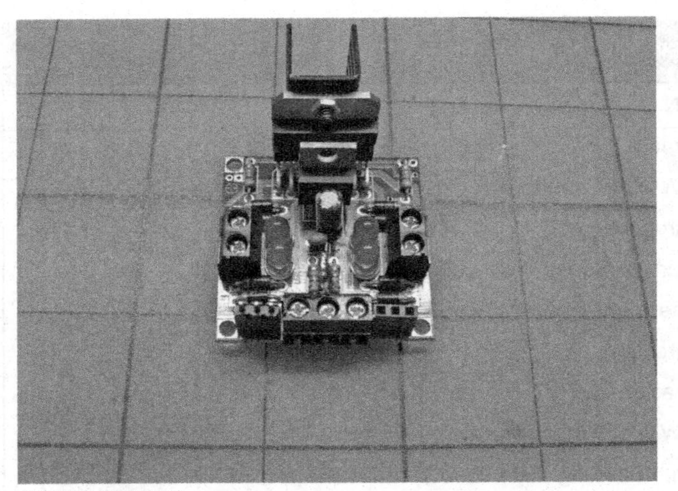

Figure 23-5 Photograph of the two-axis Solarbotics amplifier

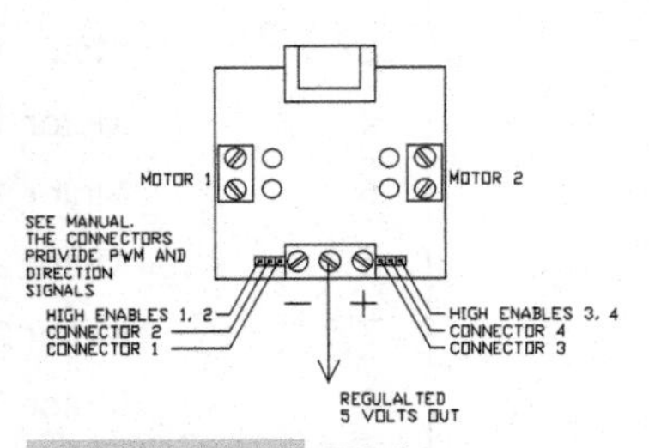

Figure 23-6 Wiring connections for two-axis Solarbotics amp

Figure 23-7 shows an example of a circuit for controlling the Solarbotics two-axis amplifier with a Propeller chip. (Note that no programs are provided for this amplifier in this book.)

Sample circuitry for using this amplifier with a microcontroller is shown in Figure 23-7.

Figure 23-8 illustrates the single-axis Xavien amplifier and Figure 23-9 illustrates how one can be controlled from a Propeller chip. (Note that no programs are provided for this amplifier in this book.)

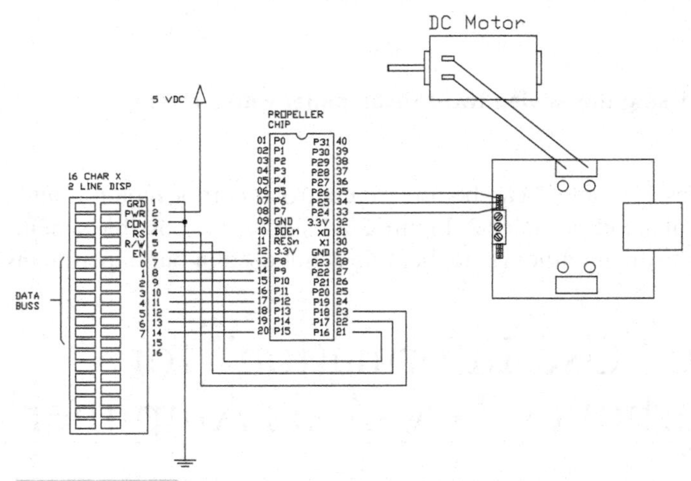

Figure 23-7 Using the Solarbotics two-axis amplifier

Figure 23-8 The Xavien single-axis amplifier

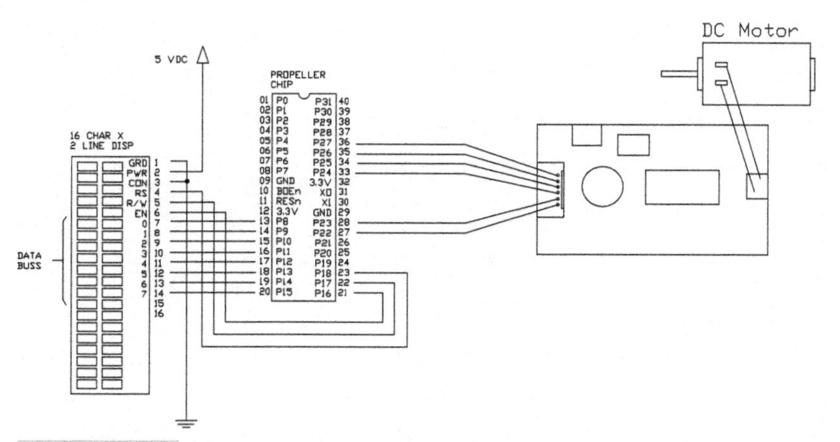

Figure 23-9 Wiring connections for Xavien single-axis amplifier

24

CONTROLLING R/C HOBBY SERVOS

The easiest way to control a simple remote operation is to use a hobby radio control transmitter, receiver, and servo to do the job. In this discussion we will consider the older 27 MHz systems as opposed to the newer GHz frequencies and the infrared systems. The older system is much more amenable to what we are doing with a Propeller running at 10 MHz and will be easier for us beginners.

Of all the various types of motors we are considering in this book, those that can be controlled by an R/C hobby radio transmitter and receiver are the simplest to use. We can readily emulate the signals that these servos need with the Propeller chip. Once the signal from the transmitter has been received and brought into the Propeller chip, we can pretty much do what we want with it before we send it out to the servo.

The servo motors have integral gearboxes built into them to allow the output shaft to move about 180 degrees from end to end. An internal potentiometer attached to the output shaft allows the system to read the position of the output shaft. Control is affected by sending the motor a signal pulse of a required duration of about 1.5 microseconds (about 60 times a second). Circuitry internal to the servo converts this to the power needed to position the servo.

R/C servos can be positioned to one part in 256 if an 8-bit variable is used to control the width of the pulse sent to them, or to one part in 4,096 if a 12-bit variable is used. For most purposes served by these small servos, the 8-bit signal is more than adequate for the job. Also, an 8-bit variable can be read into 1 byte, as compared to the 2 bytes needed for a 12-bit variable. In our particular case, we will be reading the control potentiometer to a resolution of 12 bits (with the MCP3202). Keep in mind that we can divide the value by 16 to get the 8-bit resolution when we need to.

Servo Control

The signals that come into the radio receiver are available to us at two points. One point is internal to the receiver, where the raw communications sent by the transmitter are available before they have been decoded. These signals cannot be identified or observed with ease. The other point involves the signal we get at each servo after the signal has been processed by the receiver and converted to the pulsed signal that R/C servos need to control them. A typical model aircraft servo is shown in Figure 24-1.

We will consider duplicating the servo signals after they have left the receiver and are ready to be fed to the servos. The control being implemented is the position of the servo motor. The control signal that we need to create needs to have a pulse width of 1,500 plus or minus 750 microseconds (from 750 to 2250 microseconds). This signal has to be repeated 60 times a second. The signal we create has to be mapped to whatever motor control we are interested in. In our case, the input is from a potentiometer providing a value of from 0 to 4,095, which is used to move the servo through 180 degrees. Whereas the potentiometer reading goes from 0 to 4,095, the control signal pulse width has to go from 750 to 2,250 microseconds.

MODEL AIRCRAFT SERVOS

A stated before, model aircraft servos need to have a pulsed signal sent to them approximately 60 times a second on a regular basis to accurately maintain their commanded position. If this is not done, the operation of the motors becomes jerky and irregular.

Figure 24-1 **A typical model aircraft R/C servo that uses standard Futaba wiring**

Note *This is a photograph of a standard TS-51 servo as provided by Tower Hobbies, Inc.*

This requirement also means that there can be a minimum worst-case delay of about 1/60th of a second whenever a command is sent to an R/C servo. For all practical purposes, a lag this long is not critical in a motor application; however, you need to have in mind that this delay does exist. It takes 1/60th of a second (worst case) for the motor to start responding to the last command sent to it.

Now that we know the servos have to be reminded of their position about 60 times a second in order to maintain proper operation, we need a way to pulse the servos on a regular basis 60 times a second. Because typical program flow timing is indeterminate, the pulses cannot effectively be made a part of a standard, linear program flow and still guarantee that the servos will get pulsed as needed. A better scheme is needed. In a parallel-processing system like the Propeller, this is simple. We assign one of the cogs to take care of this task.

The dedicated cog will create an overall timing cycle exactly 1/60th of a second long. Within the cycle, it will create a positive pulse between 750 and 2,250 microseconds long. The time taken by the pulse will be subtracted from the overall cycle time to ensure that the total length of the cycle is maintained at 1/60th of a second.

Wiring Connections

The standard R/C servo is a three-wire device. The signals on the three wires are as follows for the Futaba system (other systems may vary but are similar):

Wire	Color	Description
Ground	Black	Ground
Power	Red	Specified by the manufacturer (usually 5 volts will work)
Control signal	White	A TTL-level signal. This is the pulsed signal connection.

Figure 24-2 shows how the servos are wired for the three wires they need to control them.

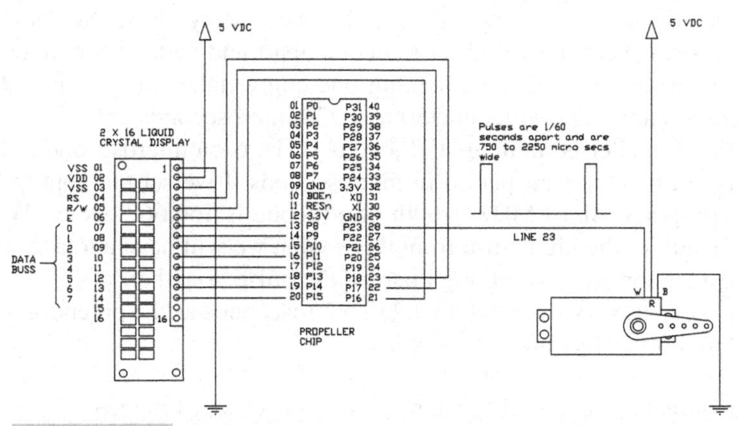

Figure 24-2 Wiring for running an R/C hobby servo from a Propeller

Note *We are using line P23 of the Propeller, but any free line can be used to control the servo.*

For the Record

- In Futaba systems, the servo center position is defined as a pulse 1.52 ms wide delivered 60 times a second. The pulse width range is 0.75 ms on either side of that. Other manufacturers specify values around 1.50 ms, so you should check exactly what your servos need in the way of the center positioning signal and the range. Also check to see that the wiring and voltage matches what you are going to provide with the Propeller. We are using the Futaba standard: white/red/black = signal/power/ground.
- Fairly large servos that follow the R/C operating standards are made for industrial usage, although they are in most probability beyond affordability for most students and hobbyists. These servos can provide adequate power for demanding laboratory and industrial applications.
- In Chapter 27 we will build a servo actuated table that uses R/C signals and two servos to control the position of the table.

DETERMINING THE POSITIONS OF THE SERVO

When we put a servo to use, it has to move to certain very specific positions to do the work we need done. We need a way to determine the exact positions needed in our applications for each servo so that we can set the positioning parameters to the appropriate values in our programs. The program we are about to create will allow you to move a servo under computer control from a potentiometer and watch the signal values that are being sent to the servo as displayed on the LCD. The setup uses a 10K potentiometer to control the position of the servo in real time. Adjust the potentiometer as needed to get the desired position for the servo and then put the values into your program for the position specified.

The code we create has to generate the pulses needed by the servo. As mentioned earlier, the pulses have to be 1/60th of a second apart and about 1,500 microseconds long. In our program, we will be assigning one cog to take care of this 1/60th of a second separation and to provide the 750 to 2,250 microseconds pulses.

Because the Propeller is running at 10,000,000 Hz, each microsecond is 10 cycles long. We can specify the wait period in microseconds if we set the appropriate constants in the program. At 10 MHz, 1/60th of a second is 166,667 cycles. We have to subtract the length of the high pulse from this so that we will have a total cycle length of 166,667 cycles for the low and high part of the pulse, together.

The pulses need to vary from 750 to 2,250 microseconds as the potentiometer value goes from 0 to 4,095. Here are the specifics:

- At a potentiometer reading of 0, we need 7,500 cycles (10 × 750).
- At a potentiometer reading of 4,095, we need 22,500 cycles (10 × 2,250).

The formula to make the conversion is

Cycles = (PotValue * (22,500 – 7,500)/4,095) + 7,500

These are the theoretical values for 10 MHz; you will have to tune this for your system once you get the program running. When we incorporate this into our Propeller program, we get the code segment shown in Program 24-1.

Program 24-1 Program Segment for Servo Control Conversion

```
outa[ServoLine]~
waitcnt((166,667-(PotValue * 15000 / 4095) + 7500 + cnt)
outa[ServoLine]~~
waitcnt((PotValue * 15000 / 4095))+ 7500 + cnt)
```

(For my particular installation, I had to use the slightly modified code to get proper servo operation. This was a function of the properties of the servo I was using.) It is almost certain that you, too, will need to modify your code to match the response of the servo you are using. Program 24-2 provides the complete listing for a program to position a servo from a potentiometer. Use this program to relate potentiometer values to actual servo positions.

Program 24-2 A “Standalone” Program for Determining the Servo Pulse Length That Matches a Specific Servo Position

```
{{ 06 Sep 09     Harprit Sandhu
RC_ServoPosition.spin
Propeller Tool Version 1.2.6
Chapter 24 Program 02

READING A SERVO'S POSITION  uses pin 23

This program allows you to tie the signal sent to an
R/C servo to the position that it goes to.

Connections

White  P23 signal from Prop
Red 5 volts
Black ground

}}
CON
  _CLKMODE=XTAL1+ PLL2X          'The system clock spec
  _XINFREQ = 5_000_000
   outPin  =23                    'line the servo is on
```

(continued)

Program 24-2 A "Standalone" Program for Determining the Servo Pulse Length That Matches a Specific Servo Position (*continued*)

```
VAR                                       'these are the variables we will use.
  long ServoPos                           'The program will read this variable from
                                          'the main Hub RAM to determine the
                                          'servo signal's high pulse duration
  long stack [15]                         'space for Cog for LCD
  long stack2[50]                         'space for Move Motor Cog
  long pcount                             '

OBJ                                        'Methods we will need for our calls
  LCD  : "LCDRoutines4"                    'for the LCD methods
  UTIL : "Utilities"                       'for general methods, pot

PUB Go                                     'main cog
  dira[outPin]~~                              'set direction for amplifier
  cognew(MoveMotor(outPin),@Stack)            'new cog and run the MoveMotor method
                                           'on it and output pulses on Pin 23
                                           'The new cog continuously reads
                                           'the "position" variable
                                           'by the example Spin code below
  cognew(cog_two, @stack2)                 'start new cog for the LCD
  repeat                                   'this is the Cog's main loop
    pcount:=UTIL.Read3202_0                'get the pot reading from the utilities
    ServoPos:=7200+pcount*4                'calculate position count

Pub cog_two                                   'set up and run the LCD
  LCD.INITIALIZE_LCD                          'initialize the LCD
  repeat                                      'LCD loop
    LCD.POSITION (1,1)                        'Go to 1st line 1st space
    LCD.PRINT(STRING("Pot Pos="))             'Potentiometer position ID
    LCD.PRINT_DEC(Pcount)                     'print the pot reading
    LCD.SPACE(2)                              'erase over old text

PUB MoveMotor(Pin)|LowTime,period             'method to toggle the output line, set
                                       'up first
  dira[Pin]~~                          'Set the direction of "Pin" to be an output
  ctra[30..26]:=%00100                 'Set this cog's "A Counter" in single
                                       'ended NCO/PWM mode (where frqa always
                                       'accumulates to phsa and the Apin output
                                       'state is bit 31 of the phsa value)
  ctra[5..0]:=Pin                      'Set the "A pin" of this cog's "A Counter"
                                       'to be "Pin"
  frqa:=1                              'Set this counter's frqa value to 1 (so 1
                                       'will be added to phsa on each clock pulse)
  ServoPos:=0                          'Start with position=0 (until the position
                                       'value is changed by another cog)
```

(*continued*)

Program 24-2 A "Standalone" Program for Determining the Servo Pulse Length That Matches a Specific Servo Position (*continued*)

```
LowTime:=clkfreq/60           'Set the time that the pulse will be low
                              'to 16ms = 1/60 secs
period:=cnt                   'Store current value of the system counter
repeat                        'line toggling routine.
  phsa:=-ServoPos             'Send a high pulse for "position" number of
                              'clock cycles. Note negative sign
  period:=period+LowTime      'Calculate what system clock's value will
                              'be at the end of this cycle's period
  waitcnt(period)             'Wait for the system counter to reach the
                              '"period" value (end of cycle)
```

We can use Program 24-2 to determine the position of a servo for use in real-world situations. It allows us to tie the reading we get from the potentiometer to the actual servo position we observe for any particular servo in our experimental setup.

In this program, we have converted the position of a potentiometer to a position of the R/C servo. In our particular case, the potentiometer was read as a value between 0 and 4,095 and the servo arm moved approximately 180 degrees. For most applications, only about 90 degrees of movement of the servo is useful. This being the case, you may want to modify the software so that the entire 4,095 values read from the potentiometer are mapped to a 90-degree movement. This will also give you finer control within the 90-degree movement.

25

CONTROLLING A SMALL DC MOTOR

For our immediate purposes, let's define a small DC motor as one that's about an inch or two in diameter and two to four inches long. We will use ones that run on about 12 volts and draw a couple of amps. The amperage and voltage values have to match the capacity of the amplifiers we have chosen for running the motors. The two-axis Xavien amplifier needs over 12 volts to operate properly and will handle 3 amps continuously and 6 amps for short bursts at up to 55 VDC for two separate motors or coils. This amplifier is used for all the experiments in the book.

Like all DC motors, the small motors shown in Figure 25-1 provide high speed and low torque. They provide no feedback in regard to the distance traveled (revolutions

Figure 25-1 **Small electric motors suitable for our experiments**

Note *These are examples of small DC motors of the type under discussion. Motors with shafts on both ends allow us to mount encoders on them for our later experiments.*

completed) or the speed of the motor. (Under certain conditions, the back EMF generated by a motor can be used as speed feedback, but we will not use this in our experiments. This use is more common with analog control schemes.)

On almost all DC motors, we can control the following parameters:

- On/off control.
- Control of power to the motor.
- Polarity of the power provided (direction of movement).
- Minimum power delivered at starting set point (power needed to start motor).
- Maximum power delivered when running as a set point. (The max RPM control depends on the load.)

Essentially, we can have comprehensive control of both the speed and the direction of these motors. Let's design a system that will give us this control of the motor, from a potentiometer and the Propeller chip/LCD system we have been using. We will design the system so that the middle position of the potentiometer will be the zero speed position, and as we turn the potentiometer in either direction, the motor will run in the selected direction. Turning the potentiometer all the way in either direction will give us full speed in the selected direction. We will provide variables in the software to control both the minimum and the maximum speed of the motor.

Potentiometer Note *In order to use a potentiometer, we need to read the position of the potentiometer wiper to get a value we can input into our control scheme. We will be using a 12-bit value for the potentiometer, so the reading will go from 0 to 4,095. We will select 2,048 as the point that gives us zero power to the motor.*

The outputs to the motor amplifier/driver will be as follows:

- An enable bit (brake=0)
- A PWM value (0 to 4,095)
- A direction bit (0 or 1)

These three values are sent to the motor amplifier/driver over the three control wires. Exactly how this is managed in the driver is a function of the motor driver we use, but most drivers do have the following three control wires for each motor:

- Enable/inhibit bit enables the driver (the Brake bit)
- PWM input for speed
- Direction bit

We can extract the direction and PWM by interpreting the 0 to 4,095 value of the potentiometer as follows (set the direction bit as detailed here):

- If the value is below 2,048, set the direction to negative. Set the direction bit to 0.
- If the value is 2,048 or above, set the direction to positive. Set direction bit to 1.

Set the speed so it will always be between 0 and 4,095:

- If the value is 2,048, set the PWM value to 0.
- If the value is above 2,048, set the PWM value to (Pot value-2048)*2.
- If the value is below 2,048, set the PWM value to (2048-Pot value)*2.

We decided earlier that we will use the larger two-axis amplifier made by Xavien. This is a very easy-to-use and fairly powerful amplifier that can handle up to 6 amps (maximum) at 55 volts (maximum) for short bursts. It accepts the three signals we need to control the motor without modification. Having two independent amplifiers on this device is a useful convenience that will allow us to use this amplifier to run our stepper motors in the various stepper experiments later in this book.

The frequency that the industry uses for the PWM signal is selected so that it is above the hearing range of humans and domestic animals. The noise is caused by loose laminations and other magnetically sensitive components in the motor. High square wave frequencies are extremely irritating to the human ear and are to be avoided. As far as the control of the motors goes, 60 Hz is completely useable (but higher frequencies do work better). We will use a frequency considerably below 20,000 Hz for our frequency because these little motors do not have a lot that will start vibrating in them at our power levels. However, do keep this in mind when you need to run a larger motor. The software we are using could easily operate at higher frequencies.

Animal Hearing Range *Most industrial amplifiers run at 40,000 Hz to keep the noise that may be generated above the hearing range of domestic animals as well.*

The circuitry needed to run our motor is shown in Figure 25-2. This circuitry reflects what needs to be wired where to run the motor with the Xavien amplifier and a Propeller chip with an attached LCD.

For power, we need two wall transformers. One to provide 7.5 to 9 volts at one amp for the Propeller power supply and one to provide between 13 and 50 volts at about one amp for the motor power supply at the amplifier. The 7.5-volt power supply provided by Parallax for the education kit is adequate for the Propeller as used in this experiment.

All the power supplies we use must be wired so that their positive terminals are at the center and the negative terminal is on the periphery of the connector. Also, 2.1-mm-diameter connectors will be used throughout. Using this standard arrangement will keep us from getting our power connections made incorrectly and thus damaging the electronics.

The connections to the Propeller will be the standard education kit layout that we always use, with the following added:

- Potentiometers
- Amplifier and motor

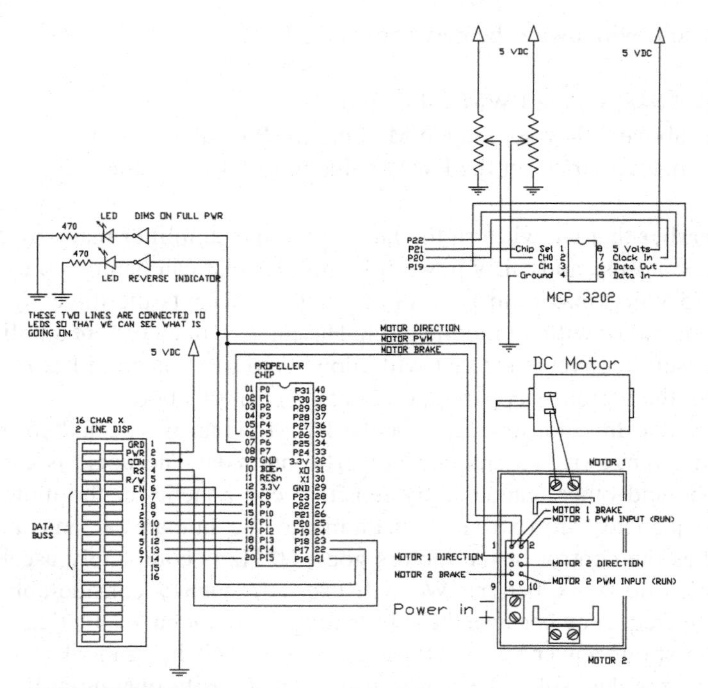

Figure 25-2 Wiring diagram for Propeller, potentiometer, and motor amplifier

- Power supplies
- LCD
- LEDs

The rest is software, so we will address that important aspect of the control next.

The Software

Running the motor in a parallel-processing environment has its own special requirements. We have to decide how to break up the goal of the total effort into various subtasks so that they can be assigned to any number of the eight cogs in a coherent way. In this experiment, we have some essentially independent tasks that are amenable to being assigned to separate cogs. A closer look at the problem reveals that we have the following separable tasks:

- Read the potentiometer that controls speed and direction.
- Generate the pulses that control the motor speed.

- Display information on the LCD so we can see what is going on.
- Turn on the two LEDs as indicators.

One way of addressing these parallel-processing needs is to manage the system so that we have the following arrangement for the cogs:

- Assign one cog to continuously read the potentiometer.
- Assign one cog to continuously handle the display functions.
- Assign one cog to determine the three motor parameters:
  - PWM value
  - Direction
  - Braking function

Doing the work in a parallel-processing environment makes it much easier to get the job done because we do not have to worry about managing a complicated interrupt-driven environment. Each of these tasks is straightforward, and we will have no problem undertaking any of them. The hard part was deciding how to divide the work up between the cogs—and for this project, even that turned out to be relatively straightforward.

A couple of the pieces of the software we need have already been developed. We already know how to read a potentiometer, and we already know how to display information on the LCD. These methods can be called from the LCDRoutines4 and Utilities programs once we have listed these objects under the OBJ block.

However, before we get ahead of ourselves, let's start with just getting the motor running. Then we will add all the other features discussed in the preceding paragraphs.

We know from the amplifier data that the amplifier is active if we tie the brake line low and make the PWM line high. The direction bit can be either high or low, and the motor will run in one direction or the other. We do, however, have to tie it high or low, because if we let it float, the motor will run back and forth in a random way, depending on the stray static electricity on the line.

After we have grounded the brake line, we need to add code for controlling the other lines. This is shown in Program 25-1. The goal here is to connect up the motor and the amplifier in the correct way and then run the program. If the motor runs, it will confirm that everything is on the right track, and we can move on to the next step.

Program 25-1 Minimal Program to Run a Motor Continuously

```
{{04 Sep 09     Harprit Sandhu
RunMotorOnly.spin
Propeller Tool Version 1.2.6
Chapter 25 Program 1
```

(continued)

Program 25-1 Minimal Program to Run a Motor Continuously (*continued*)

```
RUNNING A DC MOTOR. ON CONTROL ONLY TO
CONFIRM CONNECTIONS

This program runs the motor if turned on

Lines on Xavien amp as identified in the diagram are
line 1    Brake tie it to ground to turn it off
line 2    PWM signal     line 6 on Prop
line 3    Direction      line 7 on Prop

Connections are
Amplifier brake       P5
Amplifier PWM         P6
Amplifier direction   P7

}}
CON  _CLKMODE=XTAL1+ PLL2X                   'The system clock spec
  _XINFREQ = 5_000_000                      'crystal
  BRK = 5
  PWM = BRK + 1
  DIR = PWM + 1

PUB Go                                      'main Cog
  dira[BRK..DIR]~~                          'set direction for 3 lines
  outa[BRK..DIR]:=%000                      'make the lines low at start
   repeat
     outa[PWM]~~                            'turn on the PWM 100%
```

This program seems to be a bit on the trivial side, but it gives us one very important piece of information. It tells us that the amplifier and motor are hooked up correctly. Now, if we can manipulate the control bits under consideration with more sophistication, we will have a program like what we have in mind. If we get hung up with the software and the motor will not run, or if we think we might have blown everything up, all we have to do is run this program. If the motor runs, chances are things are still okay. At the least, it tells us that the problem is not in the hardware.

Next, let's discuss how we read the potentiometer and create the direction bit. We already know how to read a potentiometer (the Utilities object contains the methods we need). We incorporated the ability to read the potentiometer as one of the methods in the "Utilities" program. This can be done by using the Read3202_0 method as follows:

```
OBJ
       UTIL : "Utilities"

PRI
       PotentiometerCount:=UTIL.Read3202_0
```

This code places the value of the "result" value, which the Read3202_0 method created, in the PotentiometerCount variable. In the following program segment, PotentiometerCount has been shortened to Pcount to make it less cumbersome to handle. We know from the Utilities program that Pcount will be between 0 to 4,095 (in that we are reading to 12 bits with the MCP3202 chip).

We also know how to display information on the LCD, so let's incorporate all this into the next expansion of the program. The code is shown in Program 25-2. We will now be able to see the value we read from the potentiometer on the LCD, but the motor only reverses at a Pcount of 2,048 (there is no speed control).

Program 25-2 Adding the Ability to Read the Pot and Display Its Value on the LCD

```
{{04 Sep 09  Harprit Sandhu
ReadDisplayPot.spin
Propeller Tool Version 1.2.6
Chapter 25 program 2

RUNNING A MOTOR, Read Pot/NO SPEED CONTROL

This program runs the motor backward or forward
at full speed based on the potentiometer position.
Reverses at 2048
Pot position is displayed on the LCD

Lines on Xavien amp as identified in the diagram are
line 1    Brake tie it to ground to turn it off
line 2    PWM signal     line 6 on Prop
line 3    Direction      line 7 on Prop

Connections are
Amplifier brake       P5
Amplifier PWM         P6
Amplifier direction   P7

}}
CON
  _CLKMODE=XTAL1+ PLL2X          'The system clock spec
  _XINFREQ = 5_000_000           '
   BRK  =  5
   PWM  =  BRK+1
   DIR  =  BRK+2

VAR
  long position
  long stack2[50]                'space for Cog two
  long pcount                    'potentiometer reading.
   long  potpos
```

(*continued*)

Program 25-2 Adding the Ability to Read the Pot and Display Its Value on the LCD (*continued*)

```
OBJ                                   'These are the methods we will need
  LCD  : "LCDRoutines4"               'for the LCD methods
  UTIL : "Utilities"                  'for general methods

PUB Go                                'main Cog
  dira[BRK..DIR]~~                    'set direction for lines
  outa[BRK..DIR]:=%011                'make the lines high, brake off
  Cognew(cog_two, @stack2)            'start new Cog for the LCD
  repeat                              'this Cog's main loop
    pcount:=UTIL.Read3202_0           'get the pot reading from the utilities
    if pcount<2048                    'check for center position
      outa[DIR]~                      'forward direction set
      pcount:=(2048-pcount)*2         'set center,double for 4095
    else                              'or
      outa[DIR]~~                     'reverse direction
      pcount:=(pcount-2048)*2         'set center ,double for 4095
    potpos:=pcount
Pub Cog_two                           'set up and run the LCD
  LCD.INITIALIZE_LCD                  'initialize the LCD
  repeat                              'LCD loop
    LCD.POSITION (1,1)                'Go to 1st line 1st space
    LCD.PRINT(STRING("Pot position")) 'Potentiometer position ID
    LCD.POSITION (2,1)                'Go to 2nd line 1st space
    LCD.PRINT_DEC(Potpos)             'print the pot reading
    LCD.SPACE(2)                      'erase over old data
```

Once we get Program 25-2 running and the motor responding, we will have the basic on/off and reverse control of the motor in hand. We can see the position of the potentiometer on the LCD. Notice that the motor reverses at mid point on the potentiometer but still runs at full speed. We need to tie the potentiometer position/reading to the speed of the motor. In order to do this, we have to create a PWM signal whose pulse width is a function of the potentiometer reading input to the amplifier. This can be done with either of the two counters in any of the cogs in use, so next we need to learn how to use a counter. This topic was discussed earlier in the book, so you may want to review it. (See Chapter 7.) Here, we get into the details a little deeper and make it work.

A detailed description of the 31 different uses of these counters could well be a book in itself, so we will not go into explaining their use in this beginners book. We will, however, learn how to generate a PWM signal with one counter and discuss some of the general properties of the bits we manipulate to set the needed counter properties. The system assigns these bits to various functions that determine how the counter will operate. When you read up on counters in the Propeller Manual, be sure to read the sections on the frequency control registers and the phase-lock loop registers at the same time. You will have to read these sections a few times before you start to understand what is going on. The areas of interest are under instructions CTRA, FHSA, and PHSA in the Propeller Manual. It takes all three registers to control the operation of a counter.

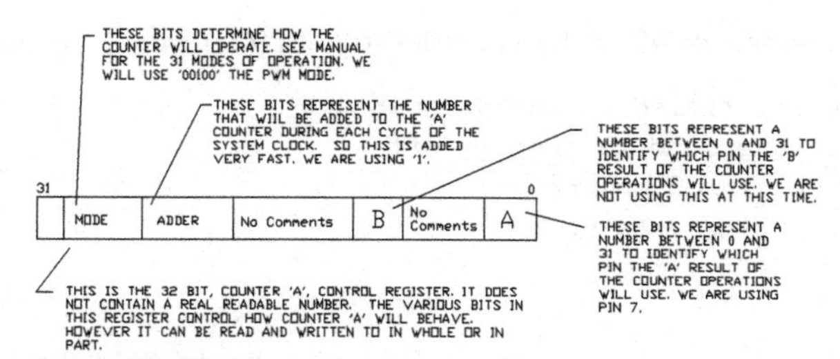

Figure 25-3 Diagram of the control register for CTRA bit assignments

Of special interest to us is the specification of the pin number that is going to provide the signal we will use as the PWM signal. This number, which can vary from 0 to 31, is specified in bits 0 to 5 in the counter control register and is called APIN (see Figure 25-3). In our case, we will be using Pin 3. If we had an interaction with another pin as a part of our selected counter operation, the number of the second pin would be specified in bits 9 to 14 and called BPIN (we do not need this pin in our present application). If we are using the phase-lock capabilities of the system, we need to specify a number that will be used for dividing the field to bring the counts down to a reasonable number. This number is specified in bits 23 to 25. These three bits can specify eight different numbers, and these numbers are listed in a table in the Propeller Manual. We will use the number 1.

There are 31 modes in which the counter can operate, and the use of all the 31 modes is beyond our expertise at this time. However, the 31 modes are specified by 5 bits in the counter. These bits go from bit 26 to bit 30. We are interested in a PWM operation. This is specified by setting these bits to 00100. (Refer to the table in the Propeller Manual.)

The duration of the phase-lock loop signal that will give us our PWM is specified in the PHSA. This is specified as a negative number so that additions to it will bring the total to 0, at which point the cycle will end and we can start over. This is the high part of the signal, the "pulse width" part of the PWM. When the low part of the signal is added into the loop, we get a complete cycle. Together, the low and high parts of the signal (the cycle time) are specified to take 10 ms in Program 25-3 (clock frequency/100).

The registers discussed in the preceding paragraphs are implemented in Program 25-3.

Program 25-3 Comprehensive DC Motor Control Program

```
{{07 Sep 09   Harprit Sandhu
MotorFromPot.spin
RUNNING A DC MOTOR FROM A POTENTIOMETER
Chapter 25 Program 3

This program allows you to control the speed and the direction of a motor
from a potentiometer.
```

(continued)

Program 25-3 Comprehensive DC Motor Control Program (*continued*)

```
Lines on Xavien amp as identified in the diagram are
line 1    Brake tie it to ground to turn it off
line 2    PWM signal     line 6 on Prop
line 3    Direction      line 7 on Prop

Connections are
Amplifier brake       P5
Amplifier PWM         P6
Amplifier direction   P7

}}
CON
  _CLKMODE=XTAL1+ PLL2X          'The system clock spec
  _XINFREQ = 5_000_000           'crystal

  BRK  =  5
  PWM  =  BRK+1
  DIR  =  BRK+2

  PotMax = 1000           'The maximum power to the motor OUT OF 4095
  PotMin =  40            'The minimum power to the motor OUT OF 4095

VAR                     'these are the variables we will use.
  long PulsWidth        'The assembly program will read this variable from
                        'the main Hub RAM to determine the
                        'servo signal's high pulse duration
  long stack [50]       'space for Cog managing the motor pulses
  long stack2[50]       'space for Cog two
  long pcount           'potentiometer reading.
  byte arrow            'motor direction arrow

OBJ                          'These are the methods we need
  LCD  : "LCDRoutines4"      'for the LCD methods
  UTIL : "Utilities"         'for general methods

PUB Go                              'main cog
  dira[BRK..DIR]~~                  'set direction for amplifier
  outa[BRK..DIR]:=%000
  cognew(MoveMotor(6),@Stack)       'start cog ,run the MoveMotor method
                                    'on it and output pulses on Pin 7
                                    'The new cog that is started above
                                    'reads the "position" variable as changed
                                    'by the example Spin code below
  cognew(cog_two, @stack2)          'start new cog for the LCD
  repeat                            'this cog's main loop
    pcount:=UTIL.Read3202_0         'get the pot reading from the utilities
    if pcount<2047                  'check for center position
```

(*continued*)

Program 25-3 Comprehensive DC Motor Control Program (*continued*)

```
      outa[DIR]~                    'forward direction set
      arrow:=0                      'direction arrow
      pcount:=(2047-pcount)*2       'set center and double
    else                            'or
       outa[DIR]~~                  'reverse direction
       arrow:=1                     'direction arrow
       pcount:=(pcount-2047)*2      'set center and double5
    if pcount<PotMin                'check Min value
       pcount:=Potmin               'set Min value
    if pcount>PotMax                'check Max value
       pcount:=Potmax               'check Max value
    PulsWidth:=pcount*24            'multiply reading to get 4095
                                    '(10000/4096=24)

PUB cog_two                           'set up and run the LCD
  LCD.INITIALIZE_LCD                  'initialize the LCD
  repeat                              'LCD loop
    LCD.POSITION (1,1)                'Go to 1st line 1st space
    LCD.PRINT(STRING("Pos="))         'Potentiometer position ID
    LCD.PRINT_DEC(pulsWidth/24)       'print the pot reading
    LCD.SPACE(2)                      'erase over old data
    LCD.POSITION (1,11)                'Go to 1st line 1st space
    LCD.PRINT(STRING("Dir="))         'Potentiometer position ID
    LCD.PRINT_DEC(arrow)              'print the pot reading
    LCD.SPACE(2)                      'erase over old data

PUB MoveMotor(Pin)|WaveLength,period    ' toggle the output line, set
                            'up first
  dira[Pin]~~               'Set the direction of "Pin" to be an output
  ctra[30..26]:=%00100      'Set this cog's "A Counter" to run in single
                            'ended NCO/PWM mode (where frqa always
                            'accumulates to phsa and the Apin output
                            'state is bit 31 of the phsa value)
  ctra[5..0]:=Pin           'Set the "A pin" of this cog's "A Counter"
                            'to be "Pin"
  frqa:=1                   'Set this counter's frqa value to 1 (so 1
                            'will be added to phsa on each clock pulse)
  PulsWidth:=1000           'Start with position=0 (until the position
                            'value is changed by another cog)
  WaveLength:=clkfreq/100   'Set the time for the pulse width to 10 ms
  period:=cnt               'Store the current value of the system counter
  repeat                    'line toggling routine.
    phsa:=-PulsWidth        'Send a high pulse for "position" number of
                            'clock cycles. NOTE minus sign on PulseWidth.
    period:=period+WaveLength 'Calculate system clock's value
                             'at the end of this cycle's period
    waitcnt(period)          'Wait for the system counter to reach the
                             '"period" value (end of cycle)
```

If you change the sign of FRQA and PHSA, you change the direction of the pulse from a going-high to a going-low pulse.

In Program 25-3, the potentiometer reading is to be handled in two ways, depending on whether the value is below 2,048 or above 2,048, and the direction bit is set accordingly. Maximum and minimum speeds are set by clamping the lowest and highest values. These values are set to 1,000 and 40 presently and do not inhibit the speed of the motor as programmed. Modify the program to set these values to 4,000 and 400, respectively, to see what happens to the motors speed.

Program 25-3 provides a comprehensive way to control a small DC motor with a Propeller chip. The program can be modified to serve the exact needs of the application you have in mind with ease and can be used as the basis for developing more sophisticated control schemes. The input does not have to be a potentiometer—any variable you can provide can be used as the controlling input. Any function that can be represented as a viable variable can be used. The point where the motor reverses can be modified, as can the rate of response of the system. Adding other decision-making algorithms to the program can lead to still further sophistication.

More importantly, a program like this does not have to control a motor at all—any task needing a PWM signal based on a variable can now be managed with ease and flexibility, and the response can be tailored to the needs of the task.

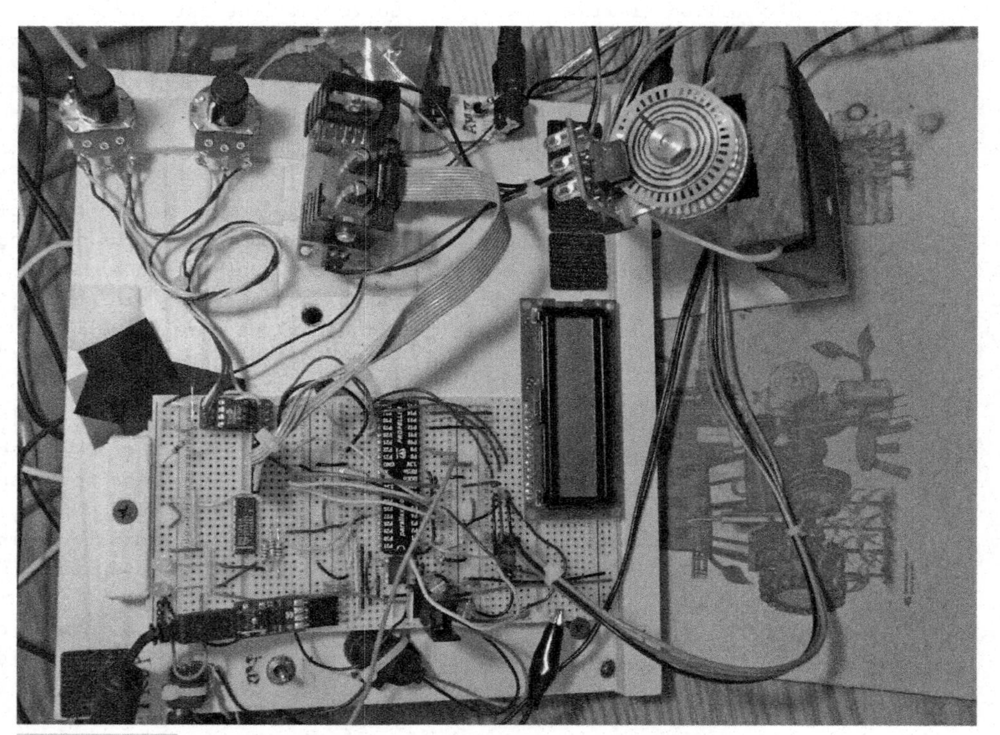

Figure 25-4 My desktop layout for the DC motor experiments

Figure 25-5 My DC motor with encoder

Figure 25-4 shows my layout of the experiments I performed to write this chapter. I did not sanitize the setting to encourage you to work with what you have. That's what it looked like when I was done! The motor I used is shown in Figure 25-5. The encoder signals were not used in this experiment but will be used later.

If you want to see what is happening in the amplifier, put the oscilloscope on the PWM line.

26

RUNNING A STEPPER MOTOR: BIPOLAR, FOUR-WIRE MOTORS

Stepper motors are made in all sorts of shapes, sizes, and configurations and for all sorts of voltages and current requirements. About the only thing they all have in common is that they move a fraction of a revolution when the electrical signals to them are changed to the next specified sequence. Because the techniques used to run all stepper motors are similar, we will concentrate on just one type of simple stepper motor for all our experiments (to keep costs down and the programming simple). Everything you learn about this one type of motor will be easily transferred to the running of other types of stepper motors. Once you know how to set up one sequence, you can set up all sequences.

In our experiments we will be using small four-wire stepper motors with only two coils inside them. These are the simplest of the stepper motors and are referred to as "bipolar" stepper motors. Bipolar motors have the added advantage of needing only one power supply. We will be using ones that needs about 12 volts at about 1 amp so that we can control them with our Xavien two-axis amplifier. Other stepper motors are similar in their needs, and once you know how to manage this bipolar motor, you should be able to run other motors without difficulty. The techniques you need to master to run this motor are the same as those used for all stepper motors.

Note *The larger Xavien two-axis amplifier, discussed earlier in the book, will handle the electrical needs of this motor with ease without need for any extra hardware. This is one of the reasons for restricting ourselves to bipolar motors. This amplifier needs a minimum of 12 VDC for the internal motor supply circuitry to work properly. A commonly available 13-volt DC wall transformer works well for this application. The voltage does not need to be closely regulated.*

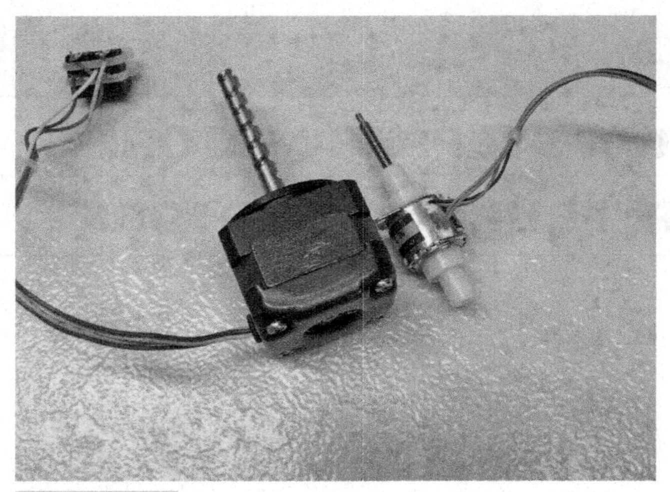

Figure 26-1 Typical small, bipolar stepper motors

Stepper Motor Power and Speed

Stepper motors provide a slow-speed and high-torque solution to our motion needs. Because they are moved an increment at a time, they also provide a means of keeping track of the amount of motion that takes place by counting the number of steps sent to the motor. As such, the most economical way to get both speed and distance motion for small projects needing limited power is to use stepper motors (see Figure 26-1).

As the designers, it is our responsibility to select a motor that will not slip or stall under load, meaning that the motor we select has to be able to carry the load without question. If the motor slips, it will not be able to respond properly to the next power sequence change, and we will lose track of the motor position. The slip/stall is considered catastrophic and cannot be recovered from unless some sort of recover/reset software is provided.

Details on Bipolar Motors

We selected bipolar stepper motors because they are the simplest of the stepper motors, they are inexpensive, and they can be run with the same amplifiers we used to run the servo motors. (The larger Xavien amplifier has two amplifiers built into it. For the servos, we needed only one of the amplifiers, but for the stepper motors we will need both, one for each coil.) A schematic representation of a bipolar stepper motor is shown in Figure 26-2.

Basically, a bipolar servo motor has a number of magnets and two sets of windings in it. When these windings are energized in the proper sequence, the motor armature rotates.

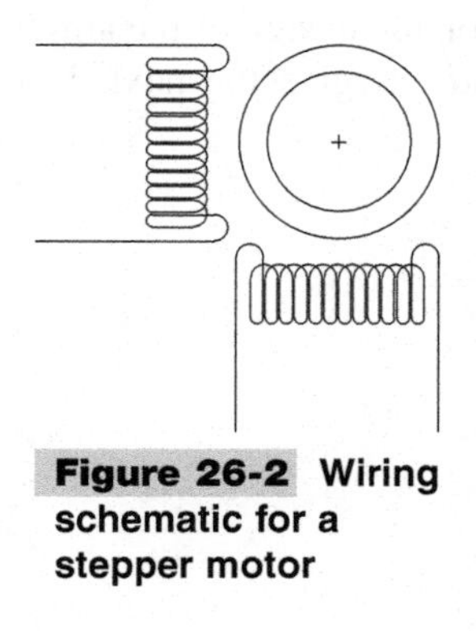

Figure 26-2 Wiring schematic for a stepper motor

The speed and direction of rotation are determined by the sequence selected and the speed at which it is executed. The steps per revolution is determined by the number of magnetic poles and matching coil arrangements in the motor; it is a property of the motor and cannot be changed.

Which two wires go together and to which winding are easily determined with a volt-ohm meter. We will connect one set of windings to one of the amplifiers and the other set to the other amplifier. As long as you do not get the coil wiring connections mixed up, it does not matter how you connect to the amplifiers because we can reverse the polarities in the software. Even so, taking an orderly approach to what we do has its advantages.

Note *This is a schematic representation and does not represent actual coil placement positions.*

Running the Motor

Once we have the motor wired to the amplifiers, we can address the business of energizing the windings in the required sequences. The sequence for movement in each direction is rigidly specified and must be followed.

Here are the only three things we can do in regard to powering up a motor winding:

- We can send current through the winding in a selected direction.
- We can reverse the direction of the current.
- We can turn the current off.

We can also modulate the current in a coil, and we can create sophisticated techniques that slowly release one coil while the other is energized. However, we will not try to create such control sequences. If you are passionate about the control of steppers, you can follow up on your own. The Propeller is powerful enough to get the job done if you care to learn its Assembly language. Spin is not fast enough.

The usual sequence for energizing the two windings can be described as follows:

1. Turn off all windings.
2. Turn on first winding (second winding off).
3. Turn on second winding; turn off first winding.
4. Reverse turn on first winding; turn off second winding.
5. Turn off second winding; reverse turn on first winding.
6. Repeat steps 2 through 5.

How the windings are turned on and off depends on the design of the amplifiers and what needs to be done to release one winding and energize the next. Here is one operations table for the Xavien amplifier:

Winding 1	**Winding 2**
ON	OFF
OFF	ON
Reverse	OFF
OFF	Reverse
Repeat	

Because a stepper motor can be programmed to be controlled in any number of ways, we have to select a few specific uses and develop the control for them. We will cover the following uses to determine the versatility of the motors and the ease of using them. Schemes will be developed to perform the following functions:

- Tie the speed of the motor to a potentiometer reading.
- Tie the distance moved to the position of a potentiometer.
- Move a motor back and forth with the motor speed based on a potentiometer.
- Move a motor back and forth with the extent of motion controlled by a potentiometer.

These basic techniques are the foundation for almost all motion required by most applications. Combining them in various ways gives you the specific results you need.

The wiring diagram for connecting a stepper motor to the Propeller chip and the potentiometer is shown in Figure 26-3.

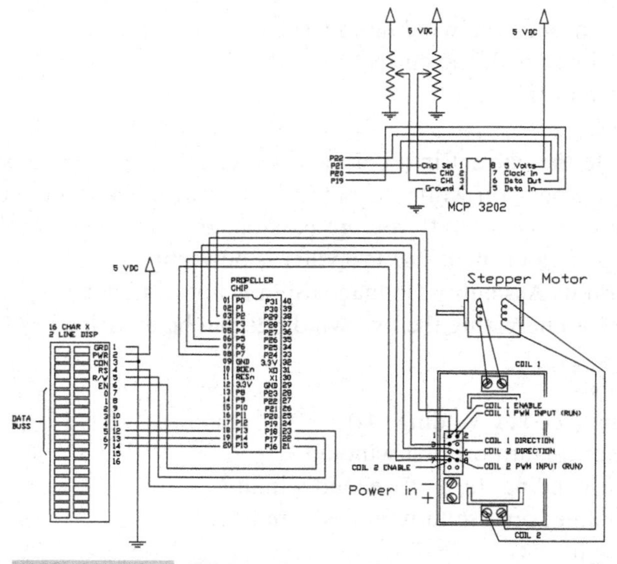

Figure 26-3 Wiring diagram for stepper motor control from potentiometers

Programming Considerations

The basic problem we need to solve involves sending the motor its control changes on a rigidly regular basis with a scheme that can still vary the time between changes without losing the regularity of the changes. Our target is to discover what the techniques are for doing this in a parallel-processing environment.

The problem cannot be solved with the usual inline programming techniques, where the program path can vary, because the time between the execution of the various instructions cannot be guaranteed and therefore neither can the programming path between subsequent motor power changes. All such techniques lead to an irregularity between consequent signals to the motor and thus to a choppy movement of the motor itself. Because of the harmonics present in a stepper motor system, this leads to problems as we increase the motor speed (the motor soon stalls or just sits there buzzing).

The usual way to solve this is to create an interrupt-based system that provides interrupts at a constant rate, where the rate of the interrupts can be controlled by the user with ease (a potentiometer in our case) and at a smooth pace. The rate at which the interrupts are generated has to meet the requirements of the lowest and highest speeds at which the motor will be expected to operate. However, we are in luck because the problem is much easier to solve in a parallel-processing environment where we do not have to contend with the difficulties associated with interrupt-based schemes.

In a parallel-processing environment, we can assign one of the processors to each critical task. In our particular case, we can break the problem into the following tasks:

- Read the input potentiometer.
- Display the results on the LCD.
- Manage the coil-energizing sequence in strictly timed steps as the timing is changed.

We have already developed the techniques for managing reading the potentiometer and displaying results on the LCD, each in its own dedicated cog. All we need to do is add a scheme for managing the coil-energizing sequences.

Obviously, the speed at which a stepper motor operates is not continuous: It is a series of time-based steps. Because the motor moves in steps, its speed is a function of the integer stepping rate that can be executed within any given time interval. Let's use one minute as our agreed-upon time interval for now. The slowest speed for a motor under these conditions will be one step per minute. If the motor is designed for 200 steps per revolution, the lowest speed at which the motor can be commanded to run is 1/200 revolutions per minute. If the maximum steps we can send it is 400 steps per second, the maximum speed is

$$(400/200)*60 \text{ rpm, or } 120 \text{ rpm}$$

We also need to be able to stop the motor, so the zero speed condition has to be mapped within the control algorithm at the lowest reading of the potentiometer (0).

If the speed is going to be controlled from a potentiometer being read into 12 bits, the 0-to-4,095 reading of the potentiometer has to be mapped to the 0 to 400 steps per second that the interrupt routine will generate. If the relationship can be made linear, that would be considered desirable for most applications.

Stepper motors also have some other characteristics you should know about before we proceed further:

- There is a limit to how fast they can be accelerated. If you try to change speed too fast, they will stutter and stall.
- There are harmonic considerations within the characteristics of the motor that make their operation at certain frequencies very smooth and at other frequencies quite problematic. They will also lose all torque at certain harmonic frequencies. Therefore, there are certain frequencies at which they cannot be run with any reliability. In some applications, it might not even be possible to pass through these frequencies for fear of stalling.
- The harmonic frequencies are affected by the load characteristics of the work being done, so the harmonic points can often be manipulated by changing the load on the motor. Both the load and the gearing of the load can be used to manipulate the harmonic points effectively.
- There is a limit to how fast a stepper motor can be run because of how fast we can create the interrupts needed and thus the rate at which the magnetic fields can be manipulated. Changing the magnetic fields in the coils back and forth causes some inductive heating. This is more severe at high frequencies. The resistive load of the coils does not change with speed but does contribute to the generation of heat within the motor.
- The torque that a stepper motor provides varies with the speed at which it is being run and is especially sensitive to collapse at critical harmonic points. Not all harmonic points behave in the same way.
- Oftentimes these handicaps can be overcome by changing the motor manufacturer or model or by changing the load or the gearing that the motor uses to drive its load.
- There are sophisticated schemes for changing the voltage and providing chopped signals to the coils to avoid some of the pitfalls as the speed is increased, but we will not consider them in this beginner's text.
- Microstepping techniques can be used to allow the motors to be moved to intermediate points between steps, but these techniques require very fast processing that can modulate the signals to the coils in real time, and we will not be able to do this with the controllers and Spin language we are using. Smoother operation between steps is also achieved by using these techniques, especially at very slow speeds. Besides, we are not limited to manipulating just one coil. Interactive coil management can lead to sophisticated, smoother control. The main focus of the more advanced techniques is to provide smoother operation and to avoid harmonic collapse with the speed range that the motor application requires.

The Software

We will be using the two-line LCD to give us feedback about what is going on within our system as we run the motor. Because the LCD is managed in its own cog, its operation does not disturb the timing of the coil signals in any way. This is one of the tremendous advantages of using a parallel-processing system. Program 26-1 lists the code segment for the LCD cog.

Program 26-1 Code Segment for LCD to Display a Message

```
Pub Cog.LCD                               'set up and run the LCD
  LCD.INITIALIZE_LCD                      'initialize the LCD
  repeat                                  'LCD loop
    LCD.POSITION (1,1)                    'Go to 1st line 1st space
    LCD.PRINT(STRING("Div=" ))            'print identifier message
    LCD.PRINT_DEC(PotValue)               'print the pot reading
    LCD.SPACE(2)                          'erase over old data
```

Note *We will be using one or two potentiometers read into 12 bits as the controlling variables for our experiments. The 0-to-4,095 readings will be used to control the range of operation of the motor in its various modes in Program 26-2 and the following programs. We have been using the Utilities program (which contains the method we need) for reading the potentiometer. The complete code was developed in Chapter 4 and is listed in Appendix A.*

For the first potentiometer, we will use

```
PotValue1:= UTIL.Read3202_0
```

And we will use the following for the second potentiometer:

```
PotValue2:= UTIL.Read3202_1
```

Program 26-2 Code Segment for Reading Potentiometers (from the Utilities Program)

```
OBJ
      UTIL : "Utilities"      'for general methods

PUB Read_Pots
   Repeat
        PotValue1:= UTIL.Read3202_0
        PotValue2:= UTIL.Read3202_1
```

What we need next is a scheme to power and reverse the windings in the right sequence for moving the motor in one direction. There are four wires on the typical bipolar motor (like the one we are using), and they are connected to the two amplifiers in the Xavien, as shown earlier in Figure 26-3.

We will be running the motor from lines P2 to P7 on the Propeller. We need three lines for each amplifier, so these six lines are adequate for the job.

The sequence for energizing the two windings is as follows, expressed in table format:

Winding 1	Winding 2
ON	OFF
OFF	ON
Reverse	OFF
OFF	Reverse
Repeat	

We are using six lines on the Propeller to control the two motor/coil amplifiers. In order to implement the on/off scheme shown in this table, the amplifier has to receive four signals. These can be managed as follows (note that the enable line is being used to turn coils on and off):

BPD	BDP	
011	010	Line service designation. B=BRK, P=PWM, D=DIR
100	101	Line P2 on the Propeller is on the left in this notation.
010	010	Line P7 on the Propeller is on the right in this notation.
100	001	

The wires from lines P2..P7 to the Xavien amplifier are connected as follows:

Propeller	Amplifier	Description
Pin P2	Pin 1	Motor 1 Brake/Enable
Pin P3	Pin 2	Motor 1 PWM
Pin P4	Pin 3	Motor 1 Direction
Pin P5	Pin 6	Motor 2 Direction
Pin P6	Pin 7	Motor 2 Brake/Enable
Pin P7	Pin 8	Motor 2 PWM

Note that in this wiring scheme, the Brake signal and the PWM signal can be used interchangeably in that each one can be used as the Enable signal or the Brake signal.

Incorporating the commands needed to reflect the connections we have made, we get the code shown in Program 26-3 for incorporation into the loop:

Program 26-3 Program Segment for Coil Power Sequence

```
repeat              'BPD DBP                where B=Brake, P=PWM, D=direction.
    outa[2..7]:=%011_010                'on  off
    waitcnt(clkfreq/divider+cnt)        'delay
    outa[2..7]:=%100_101                'off on
    waitcnt(clkfreq/divider+cnt)        'delay
    outa[2..7]:=%010_010                'rev off
    waitcnt(clkfreq/divider+cnt)        'delay
    outa[2..7]:=%100_001                'off rev
    waitcnt(clkfreq/divider+cnt)        'delay
```

The four-move loop sequence has to be repeated 25 times to get a total of 100 moves (steps). One hundred steps is one full revolution for my particular motor. The loop is repeated in the reverse direction to reverse the motor. See Program 26-4 for the code segment that has to be reversed. The "divided" value in the code segment will be read from a potentiometer so that we can change it in real time.

The preceding code segments we've developed are assembled into Program 26-4. This program provides comprehensive control of the speed of the motor in one direction from the potentiometer.

Program 26-4 Stepper Motor Speed Control from Potentiometer

```
{{09 Sep 09       Harprit Sandhu
StepperFromPot.spin
Propeller Tool Ver 1.2.6
Chapter 26 Program 4

RUNNING A STEPPER MOTOR FROM A POTENTIOMETER
UNI-DIRECTIONAL CONTROL

This program allows you to control the speed of a STEPPER motor
from a potentiometer.

 Xavien amplifier lines are
 Coil #1
   Brake is on      line P2 of the propeller
   PWM is on        line P3 of the propeller
   Direction is on line P4 of the propeller
 Coil #2
   Brake is on      line P5 of the propeller
   Direction is on line P6 of the propeller
   PWM is on        line P7 of the propeller

}}
CON
  _CLKMODE=XTAL1+ PLL2X          'The system clock spec
  _XINFREQ = 5_000_000
```

(continued)

Program 26-4 Stepper Motor Speed Control from Potentiometer (*continued*)

```
  Brk1=2
  Pwm1=Brk1+1
  Dir1=Brk1+2
  Dir2=Brk1+3
  Brk2=Brk1+4
  Pwm2=Brk1+5

VAR                         'these are the variables we will use.
  long stack1[35]           'space for Cog_LCD
  long stack2[35]           'space for Cog MoveStepper
  word Pcount               'potentiometer reading

OBJ                         'These are the methods we will need
  LCD  : "LCDRoutines4"     'for the LCD methods
  UTIL : "Utilities"        'for general methods

PUB Go                                'main cog
  cognew(MoveStepper,@Stack1)
  cognew(cog_LCD,     @stack2)        'start a new cog for the LCD
  repeat                              'this main cog's main loop
    pcount:=UTIL.read3202_0 + 1       'get the pot reading from the utilities
                                      'the 1 eliminates a 0 reading

Pub cog_LCD                              'set up and run the LCD
  LCD.INITIALIZE_LCD                     'initialize the LCD
  repeat                                 'LCD loop
    LCD.POSITION (1,1)                   'Go to 1st line 1st space
    LCD.PRINT(STRING("Div Fac=" ))       'Potentiometer position ID
    LCD.PRINT_DEC(pcount)                'print the pot reading
    LCD.SPACE(2)                         'erase over old data
    LCD.POSITION (2,1)                   'Go to 2nd line 1st space
    LCD.PRINT(STRING("Mtr Rpm=" ))       'Potentiometer position ID
    LCD.PRINT_DEC(6*pcount/10)           'print rpm
    LCD.SPACE(2)                         'erase over old data

PUB MoveStepper
  dira[Brk1..Pwm2]~~
  pcount:=50
  repeat          'BPD DBP              B=Brake, P=PWM, D=direction.
    outa[Brk1..Pwm2]:=%011_010                'on  off
    waitcnt(clkfreq/(pcount)+cnt)          '
    outa[Brk1..Pwm2]:=%100_101                'off on
    waitcnt(clkfreq/(pcount)+cnt)          '
    outa[Brk1..Pwm2]:=%010_010                'rev off
    waitcnt(clkfreq/(pcount)+cnt)          '
    outa[Brk1..Pwm2]:=%100_001                'off rev
    waitcnt(clkfreq/(pcount)+cnt)          '
```

Here are the important things to notice when you run this program: There is a limit to how fast the motor can be run, and there are some very noticeable harmonics as the motor speed is increased (the exact speeds will depend your motor). You can feel the harmonics and the loss in power as the speed increases if you put two fingers on the motor shaft and squeeze it gently as the motor is running and as you manipulate the potentiometer setting.

Next, let's modify the code so that we can control the motor direction with one potentiometer and the motor speed with another. The code is provided in Program 26-5.

Program 26-5 Pot 1 Runs Stepper Backward or Forward, and Pot 2 Controls the Speed

```
{{09 Sep 09    Harprit Sandhu
StepperFwdRev.spin
Propeller Tool Ver 1.2.6
Chapter 26 Program 5

RUNNING A STEPPER MOTOR FROM A POTENTIOMETER
Fwd-Rev control From Pot 1
Speed control from Pot 2

Xavien amplifier lines are
Coil #1
   Brake is on      line P2 of the propeller
   PWM is on        line P3 of the propeller
   Direction is on line P4 of the propeller
Coil #2
   Brake is on      line P5 of the propeller
   Direction is on line P6 of the propeller
   PWM is on        line P7 of the propeller

}}
CON
  _CLKMODE=XTAL1+ PLL2X            'The system clock spec
  _XINFREQ = 5_000_000
  Brk1=2                            'line assignments
  Pwm1=Brk1+1
  Dir1=Brk1+2
  Dir2=Brk1+3
  Brk2=Brk1+4
  Pwm2=Brk1+5
  divmin = 4
  divmax = 4500

VAR                         'these are the variables we will use.
  long stack2[25]           'space for Cog_LCD
  long stack3[25]           'space for Cog MoveStepper
  byte Pcount               'potentiometer reading.
  word divider              'divides clkfreq for delays
```

(continued)

Program 26-5 Pot 1 Runs Stepper Backward or Forward, and Pot 2 Controls the Speed (*continued*)

```
  byte direction          'stepper direction
  long range

OBJ                       'These are the methods we will need
  LCD  : "LCDRoutines4"    'for the LCD methods
  UTIL : "Utilities"       'for general methods

PUB Go                            'main Cog
  Cognew(MoveStepper,@Stack3)
  Cognew(cog_LCD,     @stack2)    'start a new Cog for the LCD
  repeat                          'this main Cog's main loop
    pcount:=UTIL.Read3202_0/16    'get the pot reading from the utilities
    range:=UTIL.Read3202_1+10     'get the pot reading for range
    if pcount<128
      pcount:=127-pcount
      direction:=0
    else
      pcount:=pcount-128
      direction:=1
    if range<divmin                 'check min
      range:=divmin                 'set min
    if range>divmax                 'check max
      range:=divmax                 'set max
    divider:=range                  'set divider value

Pub cog_LCD                         'set up and run the LCD
  LCD.INITIALIZE_LCD                'initialize the LCD
  repeat                            'LCD loop
    LCD.POSITION (1,1)              'Go to 1st line 1st space
    LCD.PRINT(STRING("Div Fac=" ))  'Potentiometer position ID
    LCD.PRINT_DEC(pcount)           'print the pot reading
    LCD.SPACE(2)                    'erase over old data
    LCD.POSITION (2,1)              'Go to 2nd line 1st space
    LCD.PRINT(STRING("Mtr Rpm=" ))  'Potentiometer position ID
    LCD.PRINT_DEC(range)            'print range/pot
    LCD.SPACE(2)                    'erase over old data

PUB MoveStepper
  dira[Brk1..Pwm2]~~
  outa[Brk1..Pwm2]~~
  divider:=500
  repeat
    if direction==0
      outa[Brk1..Pwm2]:=%011_010          'on  off
      waitcnt(clkfreq/divider+cnt)        '
```

(*continued*)

Program 26-5 Pot 1 Runs Stepper Backward or Forward, and Pot 2 Controls the Speed (*continued*)

```
    outa[Brk1..Pwm2]:=%100_001              'off rev
     waitcnt(clkfreq/divider+cnt)           '
     outa[Brk1..Pwm2]:=%010_010             'rev off
     waitcnt(clkfreq/divider+cnt)           '
     outa[Brk1..Pwm2]:=%100_101             'off on
     waitcnt(clkfreq/divider+cnt)           '
   else                                     '
     outa[Brk1..Pwm2]:=%011_010             'on  off
     waitcnt(clkfreq/divider+cnt)           '
     outa[Brk1..Pwm2]:=%100_101             'off on
     waitcnt(clkfreq/divider+cnt)           '
     outa[Brk1..Pwm2]:=%010_010             'rev off
     waitcnt(clkfreq/divider+cnt)           '
     outa[Brk1..Pwm2]:=%100_001             'off rev
     waitcnt(clkfreq/divider+cnt)           '
```

The next progression in stepper control is to convert the stepper into a servo whose position is tied to the position of the potentiometer. In order to do this, we have to keep track of how far the motor has moved from a known position and then add or subtract from that "position value" each time the motor makes a move, depending on which way the motor is moving.

The pseudocode for doing that is as follows:

```
Reset variables.
Determine potentiometer position.
Convert it to a positive or negative move equivalent (absolute position).
Move motor in direction indicated until it gets there.
Update variables as we make each step move.
```

Because we are in a parallel-processing environment, we do not have to change the methods for displaying to the LCD and reading the potentiometer, except for some minor adjustments to reflect the slightly changed needs. All the important changes are in the MoveStepper method.

The MoveStepper method is now called once for each incremental move (see Program 26-6). If we wanted, we could incorporate this into the main cog, but we will leave it in its own cog for now.

Program 26-6 Distance Moved Tied to the Position of a Potentiometer

```
{{10 Sep 09  Harprit Sandhu
StepperDistPot.spin
Propeller Tool Ver 1.2.6
Chapter 26 Program 6
```

(continued)

Program 26-6 Distance Moved Tied to the Position of a Potentiometer (*continued*)

```
RUNNING A STEPPER MOTOR FROM A POTENTIOMETER
STEPPER position FOLLOWS POT position

This program allows you to control the speed of a STEPPER motor
from a potentiometer and run motor back and forth.

Xavien amplifier lines are
 Coil #1
   Brake is on      line P2 of the propeller
   PWM is on        line P3 of the propeller
   Direction is on line P4 of the propeller
 Coil #2
   Brake is on      line P5 of the Propeller
   Direction is on line P6 of the Propeller
   PWM is on        line P7 of the propeller
}}

CON
  _CLKMODE=XTAL1+ PLL2X            'The system clock spec
  _XINFREQ = 5_000_000
  Brk1=2                           'line assignments
  Pwm1=Brk1+1
  Dir1=Brk1+2
  Dir2=Brk1+3
  Brk2=Brk1+4
  Pwm2=Brk1+5

VAR                        'these are the variables we will use.
  long stack2[35]          'space for Cog_LCD
  long stack3[35]          'space for Cog MoveStepper
  byte Pcount              'potentiometer reading.
  word divider             'divides clkfreq for delays
  byte servopos            'servo position

OBJ                        'These are the methods we will need
  LCD  : "LCDRoutines4"    'for the LCD methods
  UTIL : "Utilities"       'for general methods

PUB Go                             'main Cog
  Cognew(MoveStepper,@Stack3)
  Cognew(cog_LCD,     @stack2)     'start a new Cog for the LCD
  servopos:=32                     'any non 0 OK
  repeat                           'this main Cog's main loop
    pcount:=UTIL.Read3202_0/16     'get the pot reading from the utilities
    repeat 8                       'do 8 at a time to speed things up
      MoveStepper                  'move motor
```

(*continued*)

Program 26-6 Distance Moved Tied to the Position of a Potentiometer (*continued*)

```
Pub Cog_LCD                              'set up and run the LCD
  LCD.INITIALIZE_LCD                     'initialize the LCD
  repeat                                 'LCD loop
    LCD.POSITION (2,1)                   'Go to 2nd line 1st space
    LCD.PRINT(STRING("Target  =" ))          'Potentiometer position ID
    LCD.PRINT_DEC(pcount)                'print the pot reading
    LCD.SPACE(5)                         'erase over old data
    LCD.POSITION (1,1)                   'Go to 1st line 1st space
    LCD.PRINT(STRING("Position=" ))          'Potentiometer position ID
    LCD.PRINT_DEC(servoPos)               'print the pot reading
    LCD.SPACE(5)                          'erase over old data

Pub MoveStepper                           'stepper move routine
  dira[Brk1..Pwm2]~~                      'set at outputs
  outa[Brk1..Pwm2]~~                      'set high
  divider:=750                            'max acceptable divider
  case servopos-pcount     'this is the error in the position
     -1..-255:             'neg error move to + direction
        outa[Brk1..Pwm2]:=%011_010             'on  off
        waitcnt(clkfreq/divider+cnt)   '
        outa[Brk1..Pwm2]:=%100_101             'off on
        waitcnt(clkfreq/divider+cnt)   '
        outa[Brk1..Pwm2]:=%010_010             'rev off
        waitcnt(clkfreq/divider+cnt)   '
        outa[Brk1..Pwm2]:=%100_001             'off rev
        waitcnt(clkfreq/divider+cnt)   '
        servoPos:=servoPos+1
     0:                    'no error so do nothing
     1..255:               'pos error so move in- direction
        outa[Brk1..Pwm2]:=%011_010             'on  off
        waitcnt(clkfreq/divider+cnt)   '
        outa[Brk1..Pwm2]:=%100_001             'off rev
        waitcnt(clkfreq/divider+cnt)   '
        outa[Brk1..Pwm2]:=%010_010             'rev off
        waitcnt(clkfreq/divider+cnt)   '
        outa[Brk1..Pwm2]:=%100_101             'off on
        waitcnt(clkfreq/divider+cnt)   '
        servoPos:=servoPos-1
```

In Program 26-6, we have to slow the servo stepping time down and then reduce the range of the potentiometer values to give us an approximately 16-revolution movement of the servo from one extreme to the other. Manipulate these factors to see what happens so you can understand why these compromises have to be made.

One often-used stepper scheme moves a motor back and forth constantly to perform a repetitive task. An example of this would be the movement of a printer head back and forth across the paper to access an entire page. We will develop the code to do this in the Program 26-8. We will fix the extent of the movement to about 16 revolutions, and adjust the speed with the potentiometer to see what happens.

A very interesting phenomenon becomes evident when you have to reverse the motor and start from zero speed again and again: Although the motor can run at a high speed if we ramp up to the speed, it cannot start at a high speed setting. In other words, we have to ramp up every time. Program 26-8 uses the code segment from Program 26-7 to control the ramp up.

Program 26-7 Program Segment to Ramp Up Stepper Speed

```
PUB MoveStepper
  dira[BRK..DIR]~~
  outa[BRK..DIR]~~
  divider:=divstt ''  'divider start setting, a constant
  repeat              'the outside loop for fwd and back
    repeat pcount     'the forward part of the outer loop
      divider:=divider*divstt/(divstt-20)  'increase speed in each loop
      if divider>divmax'           'check that you don't exceed maximum
        divider:=divmax                'clamp at max speed
      outa[2..7]:=%011_010             'on  off
      waitcnt(clkfreq/divider+cnt)     '
      outa[2..7]:=%100_101             'off on
      waitcnt(clkfreq/divider+cnt)     '
      outa[2..7]:=%010_010             'rev off
      waitcnt(clkfreq/divider+cnt)     '
      outa[2..7]:=%100_001             'off rev
      waitcnt(clkfreq/divider+cnt)     '
   divider:=divstt                     'reset the divider
   waitcnt(clkfreq/2+cnt)              'pause to look at LCD
```

In this program segment, the divider divides the system clock by a number to determine the time between stepper moves. Each time through the loop of four moves, the value of the divider is increased, thus decreasing the time between moves. This could have been done between each move to act even faster, but I chose not to, to keep the code simple and more compact.

The program then checks to make sure we have not made the time too short by clamping the divider to divmax. Both divstt and divmax are constants set at the top of the program. Play with these to see what happens and then figure out what to do about it. The numbers for your stepper, most probably, will be different from the numbers I have used in the sample programs. The code that demonstrates this is given in Program 26-8.

Program 26-8 Move a Stepper Motor Back and Forth with Distance and Motor Speed Controlled by Potentiometers

```
{{10 Sep 09  Harprit Sandhu
StepBackForthPot.spin
Propeller Tool Ver 1.2.6
Chapter 26 Program 8
```

(continued)

Program 26-8 Move a Stepper Motor Back and Forth with Distance and Motor Speed Controlled by Potentiometers (*continued*)

```
RUNNING A STEPPER MOTOR FROM A POTENTIOMETER
MOVES BACK AND FORTH
Pot 1 controls distance
Pot 2 controls max speed.

Xavien amplifier lines are same as before
 Coil #1
   Brake is on      line P2 of the propeller
   PWM is on        line P3 of the propeller
   Direction is on line P4 of the propeller
 Coil #2
   Brake is on      line P5 of the Propeller
   Direction is on line P6 of the Propeller
   PWM is on        line P7 of the propeller

}}
CON
  _CLKMODE=XTAL1+ PLL2X          'The system clock spec
  _XINFREQ = 5_000_000
  Brk1=2                          'line assignments
  Pwm1=Brk1+1
  Dir1=Brk1+2
  Dir2=Brk1+3
  Brk2=Brk1+4
  Pwm2=Brk1+5
  divstt  = 200            'starting speed

VAR                        'these are the variables we will use.
  long stack2[25]          'space for Cog_LCD
  long stack3[25]          'space for Cog MoveStepper
  word Pcount              'potentiometer reading.
  word divider             'divides clkfreq for delays
  word divmax              '

OBJ                        'These are the methods we will need
  LCD  : "LCDRoutines4"     'for the LCD methods
  UTIL : "Utilities"       'for general methods

PUB Go                            'main Cog
  Cognew(MoveStepper,@Stack3)     'start new Cog for stepper moves
  Cognew(cog_LCD,    @stack2)     'start a new Cog for the LCD
  repeat                          'this main Cog's main loop
    pcount:=UTIL.Read3202_0/4     '
    divmax:=100*UTIL.Read3202_1*2/82+2    'max is 10000
```

(*continued*)

Program 26-8 Move a Stepper Motor Back and Forth with Distance and Motor Speed Controlled by Potentiometers (*continued*)

```
Pub Cog_LCD                              'set up and run the LCD
  LCD.INITIALIZE_LCD                     'initialize the LCD
  repeat                                 'LCD loop
    LCD.POSITION (1,1)                   'Go to 1st line 1st space
    LCD.PRINT(STRING("Div =" ))          'Potentiometer position ID
    LCD.PRINT_DEC(divider)               'print the pot reading
    LCD.SPACE(2)                         'erase over old data
    LCD.POSITION (1,12)                  'Go to 1st line 12th space
    LCD.PRINT_DEC(divmax)                'max division
    LCD.SPACE(2)                         'erase over old data
    LCD.POSITION (2,1)                   'Go to 2nd line 1st space
    LCD.PRINT(STRING("Dist=" ))          'Potentiometer position ID
    LCD.PRINT_DEC(pcount)                'print the pot reading
    LCD.SPACE(2)                         'erase over old data
    LCD.POSITION (2,12)                  'Go to 2nd line 12th space
    LCD.PRINT(STRING("Max"))             'max division
    LCD.SPACE(2)                         'erase over old data

PUB MoveStepper
  dira[BRK1..PWM2]~~
  outa[BRK1..PWM2]~~
  divider:=divstt
  repeat
    repeat pcount
      divider:=divider*divstt/(divstt-10)
      if divider>divmax
        divider:=divmax                  'BPD DBP B=Brake,P=PWM,D=direction.
      outa[BRK1..PWM2]:=%011_010               'on  off
      waitcnt(clkfreq/divider+cnt)       '
      outa[BRK1..PWM2]:=%100_101               'off on
      waitcnt(clkfreq/divider+cnt)       '
      outa[BRK1..PWM2]:=%010_010               'rev off
      waitcnt(clkfreq/divider+cnt)       '
      outa[BRK1..PWM2]:=%100_001               'off rev
      waitcnt(clkfreq/divider+cnt)    '
   divider:=divstt
   waitcnt(clkfreq/3+cnt)

    repeat pcount
      divider:=divider*divstt/(divstt-10)
      if divider>divmax
        divider:=divmax
      outa[BRK1..PWM2]:=%011_010               'on  off
      waitcnt(clkfreq/divider+cnt)       '
      outa[BRK1..PWM2]:=%100_001               'off on
```

(*continued*)

Program 26-8 Move a Stepper Motor Back and Forth with Distance and Motor Speed Controlled by Potentiometers (*continued*)

```
      waitcnt(clkfreq/divider+cnt)       '
      outa[BRK1..PWM2]:=%010_010                'rev off
      waitcnt(clkfreq/divider+cnt)       '
      outa[BRK1..PWM2]:=%100_101                'off rev
      waitcnt(clkfreq/divider+cnt)  '
   divider:=divstt
   waitcnt(clkfreq/3+cnt)
```

We discover that one more important thing needs to be controlled. When the motor stops, it has to either hold its position or be allowed to freewheel. The last instruction has to be designed to manage this aspect of the motor operation in a way that will not overheat the motor over a long period of time. One way would be to reduce the current to the motor when on standby, with a separate method, using a suitable PWM technique that reduces the energy supplied to the motor to an acceptable level.

A Fine Point about Stepper Control *The motor-powering scheme follows a 1, 2, 3, 4 sequence. When a motor stops, it must restart with a valid next sequence. For example, when the motor stops after executing the fourth sequence, the next move must be a 3 or a 1 sequence; it cannot be a 2 sequence. A 2 is not valid after 4. If this scheme is not followed, the motor can lose counts and the first move can be unpredictable or hesitant. You can see this problem in Program 26-8 if you watch carefully. See if you can fix it.*

By now we should have a reasonably good feel for what it takes to control a stepper motor with a Propeller chip and the Spin language. Let's go ahead and put all we have learned into a program that lets us control the distance moved by the motor with a potentiometer. As before, the maximum speed for each move is specified by the divmax constant at the top of the program. Program 26-9 is similar to what we developed in Program 26-8 and is optimized for the stepper motor I used. You will have to fine-tune this program for your particular motor and load.

Program 26-9 Move a Motor Back and Forth with the Extent of Motion Controlled by a Potentiometer

```
{{24 Sep 09     Harprit Sandhu
StepFromPot.spin
Propeller Tool Ver. 1.2.6
Chapter 26 Program 9

Running a stepper motor from a potentiometer
Moves back and forth a variable distance
Max speed fixed but can be changed at divmax constant
Control distance moved with the potentiometer.
Finished program
```

(*continued*)

Program 26-9 Move a Motor Back and Forth with the Extent of Motion Controlled by a Potentiometer (*continued*)

```
This program allows you to control the distance moved by a STEPPER
motor from a potentiometer as it moves back and forth automatically
The number of steps it goes up and down is determined by the pot
and is initially set for a maximum move of ~100 turns for a motor
with 100 steps per revolution (which I was using).
Display shows the divider value used and Pot reading

 Xavien amplifier lines
 Coil #1
   Brake is on     line P2 of the propeller
   PWM is on       line P3 of the propeller
   Direction is on line P4 of the propeller
 Coil #2
   Brake is on     line P5 of the Propeller
   Direction is on line P6 of the Propeller
   PWM is on       line P7 of the propeller

}}
CON
  _CLKMODE=XTAL1+ PLL2X 'The system clock spec
  _XINFREQ     =5_000_000
  divmax       =8000          'the higher this is the faster the speed
  divstart     =200           'starting speed
  divDecrement =20            'rate at which speed changes

VAR                          'these are the variables we will use.
  long stack1[35]            'space for Cog_LCD
  long stack2[35]            'space for Cog MoveStepper
  word Pcount                'potentiometer reading.
  word divider               'divides clkfreq for delays

OBJ                          'These are the methods we will need
  LCD  : "LCDRoutines4"      'for the LCD methods
  UTIL : "Utilities"         'for general methods

PUB Go                              'main Cog
  Cognew(cog_LCD,     @stack1)      'start a new Cog for the LCD
  Cognew(MoveStepper,@Stack2)       'start new Cog for stepper moves
  repeat                            'this is main Cogs main loop
    pcount:=UTIL.read3202_0/4+1     'get the pot reading from the utilities

PRI Cog_LCD                            'set up and run the LCD
  LCD.INITIALIZE_LCD                   'initialize the LCD
  repeat                               'LCD loop
    LCD.POSITION (1,1)                 'Go to 1st line 1st space
    LCD.PRINT(STRING("Div =" ))        'dividing value
```

(continued)

Program 26-9 Move a Motor Back and Forth with the Extent of Motion Controlled by a Potentiometer (*continued*)

```
    LCD.PRINT_DEC(divider)          'print value
    LCD.SPACE(5)                    'erase over old data
    LCD.POSITION (2,1)              'Go to 2nd line 1st space
    LCD.PRINT(STRING("Dist=" ))     'Distance moved
    LCD.PRINT_DEC(pcount)           'print value
    LCD.SPACE(5)                    'erase over old data

PRI MoveStepper                     '
  dira[2..7]~~                      'all outputs
  outa[2..7]~~                      'all high
  divider:=divstart                 'set starting point for div
  repeat                            'outer loop
    repeat pcount                   'move loop, forward
      divider:=divider*divstart/(divstart-divDecrement)
      if divider>divmax                 'check div
        divider:=divmax                 'clamp div
      outa[2..7]:=%011_010              'on  off
      waitcnt(clkfreq/divider+cnt)      '
      outa[2..7]:=%100_101              'off on
      waitcnt(clkfreq/divider+cnt)      '
      outa[2..7]:=%010_010              'rev off
      waitcnt(clkfreq/divider+cnt)      '
      outa[2..7]:=%100_001              'off rev
      waitcnt(clkfreq/divider+cnt)      '
    divider:=divstart                    'reset divider
    waitcnt(clkfreq/8+cnt)

    repeat pcount                       'move loop, backwards
      divider:=divider*divstart/(divstart-divDecrement)
      if divider>divmax                 'check div
        divider:=divmax                 'clamp div
      outa[2..7]:=%011_010              'on  off
      waitcnt(clkfreq/divider+cnt)      '
      outa[2..7]:=%100_001              'off on
      waitcnt(clkfreq/divider+cnt)      '
      outa[2..7]:=%010_010              'rev off
      waitcnt(clkfreq/divider+cnt)      '
      outa[2..7]:=%100_101              'off rev
      waitcnt(clkfreq/divider+cnt)
    divider:=divstart                   'reset divider
    waitcnt(clkfreq/8+cnt)
```

The techniques that have been demonstrated can be combined with one another to provide the stepper control you would need for almost any application. The keys are proper motor selection and proper ramping. And making sure the system never stalls or slips.

27

GRAVITY SENSOR BASED AUTO-LEVELING TABLE

The simplest of construction is adequate for our purposes. The important thing is that you actually build a table. My table was made with cardboard and hot glue and bamboo skewers for the bearing shafts.

In this project, we will use the Memsic 2125 (which was covered in Chapter 18) to build a table that stays horizontal while the base of the table is disturbed (within about 20 degrees) relative to the horizon.

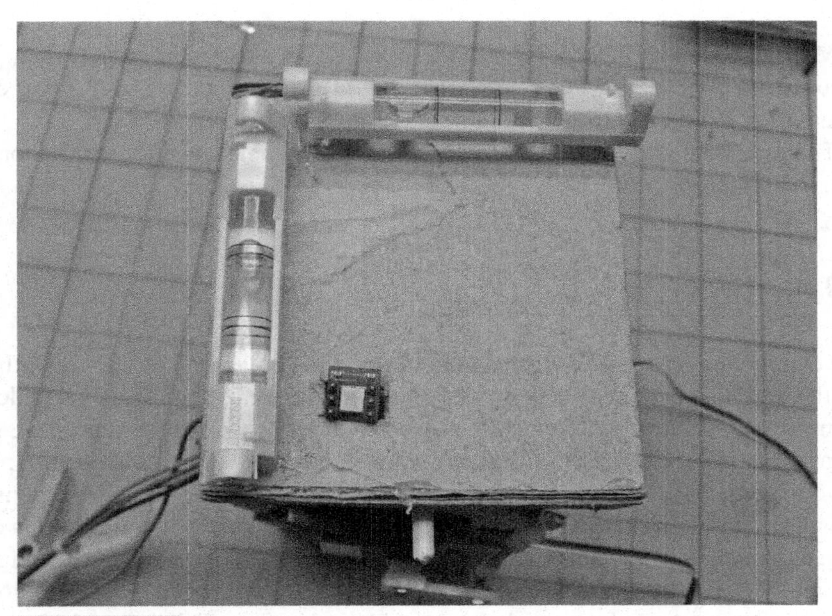

Figure 27-1 An artificial horizon—a basic unadorned cardboard horizontal table with two plastic levels and a Memsic 2125 sensor

Sensor Specifications

Here is more detailed information about the Memsic accelerometer (from Parallax):

> "The Memsic 2125 is a low-cost, dual-axis thermal accelerometer capable of measuring dynamic acceleration (vibration) and static acceleration (gravity) with a range of ±2 g. For integration into existing applications, the Memsic 2125 is electrically compatible with other popular accelerometers."
>
> Key features of the Memsic 2125 include the following:

- Measure 0 to ±2 g on either axis; less than 1 mg resolution
- Fully temperature compensated over 0°C to 70°C range
- Simple, pulse output of G-force for X and Y axes
- Analog output of temperature (T-Out pin)
- Low current operation: less than 4 mA at 5 VDC

A sampling of possible BASIC Stamp module applications with the Memsic 2125 include:

- Dual-axis tilt sensing for autonomous robotics applications
- Single-axis rotational position sensing
- Movement/lack-of-movement sensing for alarm systems
- R/C hobby projects such as autopilots

Memsic (www.memsic.com) provides the 2125 in a surface-mount format. Parallax mounts the circuit on a PC board providing all I/O connections, so it can easily be inserted on a breadboard or through-hole prototype area.

The actuators we use to move the table for horizontal position correction will be two radio control hobby servos.

Discussion

We can mount the sensor to correct the tilt in two locations: We can mount the sensor on the base that is being tilted, or we can mount the sensor on the actual table that we want to keep level. If we mount the sensor on the base we will be tilting, we get error signals that we can interpret to make corrections to the target table position. We may have to create a lookup table if there is any nonlinearity either in the response of the sensor or in the mechanical linkages that connect the base to the top. In this case, the signal that we detect is absolutely related to the position of the base, meaning that we are reading the actual error at the base. What we do with the signal is up to us, and how we design the linkages to level the table is up to us. This is not the best method to use in most cases, but this may be the only way available to us in some situations.

If we mount the sensor to the table top itself, the signal we get will be a measure of how far the table has tilted. We can make the table come back to horizontal if we keep making a correction until the table becomes horizontal, and we stop when the error signal goes to zero. It is a process integrated over time. (If we wanted to know how large the correction was, we would have to keep track of how far we had moved the table to get it back to horizontal, but we do not need this information in an integrating system.) We could also relate the error signal directly to the amount of correction needed and bring the table back to level in that way, but the bookkeeping required to do that is complicated.

Simply stated, it is better to mount the detector to the actual tabletop because that is the surface we are interested in keeping level. As a general rule, the closer the sensor is to the actual error-signal-creating element, the easier it will be to know what the error signal means and the easier it will be to make a correction.

SETTING UP THE HARDWARE CONNECTIONS

The system we will create needs to read two inputs from the sensor and output two outputs, one to each servo. The inputs are in the form of two pulse widths that we will read from the X-Out and Y-Out connectors of the sensor, and our outputs from the Propeller are the two 1- to 2-millisecond pulses we will send to the radio control hobby servos 60 times a second. See Figure 27-2 for wiring details.

Note *There are four wires between the sensor and the system, and both sides of Memsic must be grounded.*

We investigated the Memsic in Chapter 18. Review that chapter now if you need to. We will base our software in this chapter on the software created in Chapter 18, with additions and modifications.

The Memsic puts out a fixed-frequency pulse with a variable duty cycle. On my sensor, I received a wave length of almost exactly 100,000 cycles, which is a frequency of

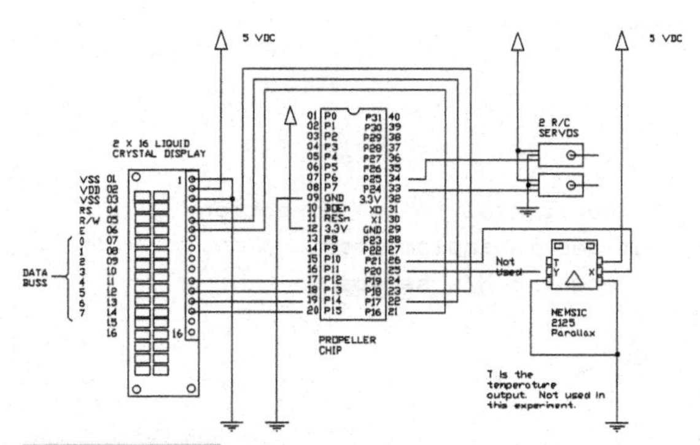

Figure 27-2 **Wiring diagram for the table: connecting to the Memsic accelerometer to two servos**

100 Hz. At level, the duty cycle of the PWM signal was 50%, or 50,000 cycles. (As always, my system was running at 10 MHz.)

The Futaba-compatible servos we are using require a center position pulse of 1,520 microseconds bracketed with a range of + or –750 microseconds (you have to check this on your specific servos). At 10 MHz, this is 15,200 + or –7,500 cycles.

Here's the equation for converting what we read into what the servos need:

Output pulse length = 1,520 + (reading – 50,000)/10 microseconds

We can implement these conditions with the following pseudocode for the operation of one of the axes on the table (you can expand it for full two-axis operation when you build your table):

```
Initialize the system.
Define all parameters and lines to be used.
Read the X axis on the Memsic sensor.
Compare it to the horizontal position.
CASE
    If it is positive, move in the negative direction one step.
    If there is no error, do nothing.
    If it is negative, move in the positive direction one step.
Repeat to read sensor.
```

Advanced Options *If you want to be able to adjust the position of the table with respect to the true horizon, you can add two potentiometers to the hardware to act as final correction inputs. You may want to use the LCD to provide information about the conditions in the system during development and later to annunciate the system status.*

The software that implements single-axis operation defined by the pseudocode is described in Program 27-1.

Program 27-1 Program for Single-Axis Correction to Horizon Table

```
{{04 Jan 10     Harprit Sandhu
MemsicTable.spin
Propeller Tool Ver 1.2.6
Program 27-1

This program keeps the table horizontal in one direction.
Both servos are connected but one is implemented.
It uses a Memsic 2125 sensor and two R/C servos.

COG_LCD manages the LCD output
COG_0 measures the pulse
COG_1 manages the servo

}}
```

(continued)

Program 27-1 Program for Single-Axis Correction to Horizon Table (*continued*)

```
CON
  _CLKMODE=XTAL1+ PLL2X        'The system clock spec
  _XINFREQ = 5_000_000         'crystal frequency

  Xaxis  = 26                'from Memsic
  Yaxis  = 27                'for speaker

  chipSel  = 19                'for pots
  chipClk  = chipSel+1         '
  chipDout = chipSel+2         '
  chipDin  = chipSel+3         '

  servo    = 23                'to servo signal
  servo2   = 24                'to servo2 signal

VAR
  long  Stack[50]                'FOR LCD COG
  long  Stack1[50]               'FOR SERVO COG
  long  startWave                '
  long  endPulse                 '
  long  endWave                  '
  long  PulseLen                 '
  long  waveLen                  '
  long  frequency                '
  long  servoPos                 '

OBJ                              'These are the Objects we will need
  LCD  : "LCDRoutines4"          'for controlling the LCD
  UTIL : "Utilities"             'for general methods collection

PUB go                             'Cog_0
  cognew (COG_LCD, @Stack)         'starting up Cog LCD
  cognew (SERVO1, @Stack1)         'starting up Cog OUT
  DIRA[25]~                        'Make pin input
  repeat                               'Set up the control read loop
    repeat while ina[xaxis]==1         'wait for line 1 to go hi. See Manual
    repeat while ina[xaxis]==0         'wait for line 1 to go low. Manual
    startWave:=cnt                     'read the timer count
    repeat while ina[xaxis]==1         'wait for line 1 to go hi. See Manual
    endPulse:=cnt                      'read the timer count for second time
    repeat while ina[xaxis]==0         'wait for line 1 to go low. Manual.
    endWave:=cnt                       'end of wave cycle
    PulseLen:=endPulse-startWave       'figure the pulse
    waveLen:=endWave-startWave         'figure the wave Len
    frequency:=clkfreq/waveLen         'figure the freq
```

(*continued*)

Program 27-1 Program for Single-Axis Correction to Horizon Table (*continued*)

```
PRI COG_LCD                         'This is running in the new cog
  LCD.INITIALIZE_LCD                'set up the LCD
  repeat                            'LCD routine loop
    LCD.POSITION (1,1)              'Position LCD cursor
    LCD.PRINT(String("PL="))        'Pulse
    LCD.PRINT_DEC((pulselen))       'print data
    LCD.SPACE(2)                    'write over old data
    LCD.PRINT_DEC((servoPos))       'print servo position
    LCD.SPACE(2)                    'write over old data
    LCD.POSITION (2,1)              'Position LCD cursor
    LCD.PRINT(String("WL="))        'Wave Length
    LCD.PRINT_DEC((wavelen))        'print value
    LCD.SPACE(2)                    'write over old data
    LCD.POSITION (2,11)             'Position LCD cursor
    LCD.PRINT(String("FR="))        'Frequency
    LCD.PRINT_DEC((frequency))      'print value
    LCD.SPACE(2)                    'write over old data

PRI SERVO1                          'servo positioning routine
dira[servo]~~                       'set pin direction
servoPos:=15000                     'initial position
  repeat
    case pulseLen                               'based on servo position
      0..50700:servoPos:=servoPos+200           'move positive
      50701..50800:                             'no move needed
      50801..100_000:servoPos:=servoPos-200     'move negative
    servoPos #>=6500                            'limit to 6500
    servoPos <#=22500                           'limit to 22500
    outa[servo]~~                               'send out servo pulse
    waitcnt(servoPos+cnt)                       'pulse length
    outa[servo]~                                'end pulse
    waitcnt(clkfreq/60+cnt)                     '60 per second pause.
```

In this program, the 200 in the SERVO1 method determines the sensitivity of the response. It is multiplying what is essentially the error signal. Try changing its value to 500 and see what happens. Try 100 and then try 10.

The at-rest (horizontal) reading from the Memsic on my particular sensor is 50700. There may be slight variations in the value from sensor to sensor, so we may want to add a potentiometer to the circuit wiring so we can make an adjustment (say, to always bring this value to 50700). On the servo I was using, the center position was 1320 microseconds based on how the horn was attached to the servo. The theoretical value for this is usually stated as 1,520 microseconds (by Futaba). Each servo can be expected to be slightly different, and the mounting position of the servo arm/horn to the servo also affects this number. Here again, we could provide a trim potentiometer to fine-tune this value.

The preceding exercise demonstrates the relative ease with which we can make a fairly sophisticated instrument, such as an artificial horizon, when we use a Propeller microcontroller.

This instrument could easily be modified to show how many G-forces one went through as one turned a corner in a car. In this case, the sensing axis of the sensor would have to be placed left to right across the automobile and a multiplier would have to be adjusted to give a reasonable display on the LCD.

TWO-AXIS SOFTWARE

To make the servo table maintain horizontal position in both directions, we have to add the code for the second servo. Because this is essentially a repeat of the single-axis operation, this is left up to you. Once both axes are operational, see what needs to be done to make the operation of the table as crisp and responsive as possible.

> **Hint** *We know there is a 1/60th of a second worst-case delay in the operation of the servos that will limit the speed of the response, but other things may be possible, such as moving in larger steps to correct the positional errors and creating a totally new algorithm for the correction.*

BUILDING THE ARTIFICIAL HORIZON TABLE

I cannot overemphasize the fact that we learn best by doing. Therefore, it is important that you actually build a table because you will discover all kinds of things about its operation that I cannot possibly convey in words. I have made it as simple and as inexpensive as I could for you to be able to undertake this exercise. Do it!

The easiest way for us to construct the table is to build two tables, one above the other, with the two controlled axes at right angles to one another. The major problem is that the two axes will not be completely independent because of the mechanical interactions in a simple—and not very precise—cardboard construction. Because the correcting mechanism designed for each table setup will be different, the implementation thereof will be left up to each individual constructor. However, we have covered the general principals. Using an integrating scheme to make the corrections eliminates the need for absolute corrections.

Figure 27-3 shows a table I built out of cardboard. The table components were cut out from an old cardboard box with a box knife and a pair of scissors and then put together with hot glue. Because our interest is in the control of the table as opposed to the table itself, a simple cardboard model allows us to do what we want at minimal cost and with minimal effort. Fortunately, the slight sloppiness and inaccuracies in the construction of the table lend themselves to software corrections.

After the servos have been added to the table and the sensor placed on the top surface, the table mechanism then looks like what is shown in Figure 27-4.

Figure 27-3 Picture of a simple horizontal table and the gravity sensor.

Note *The basic table mechanism can be made out of cardboard.*

Note *Mechanical inaccuracies will be compensated for by the software.*

The Memsic detector we are using needs two wires between the sensor and the Propeller, a power connection, and ground. A simple extension cable with a four-point female or male end can be made to fit to the breadboard. Because we are using

Figure 27-4 **Picture of the cardboard table mechanism with servos installed.**

a relatively expensive sensor, we will not want to solder to it. Therefore, suitable slip-on connections should be provided on the sensor end also.

It is left up to you to modify the software to increase the speed with which the table returns to horizontal after the base has been moved. Rather crisp operation is possible with a little imagination.

28

RUNNING DC MOTORS WITH ATTACHED INCREMENTAL ENCODERS

If you are not yet quite comfortable with the parallel Propeller system, you may want to wait until you are before you address this next to last chapter in the book. It is important that you understand each and every line of code in all these programs. There are subtle changes in the successive programs, and I have tried to call them out for you, but you still need to go over each program a line at a time and understand it.

These small motors have simple two-phase quadrature encoders attached to them. They are available at Encodergeek.com (see Figure 28-1).

Figure 28-1 Small DC electric motors with encoders

Not about Motors

This is the most sophisticated program development in the book, and you need to understand up front that it is not really about running motors. *It is about understanding any control system with some form of feedback,* a very important skill to develop. In this case, the feedback is coming from an optical encoder, but it could be anything. Even though we are controlling a motor, it could be any control point of interest to us. A motor has been picked because this is a system that is easy to purchase and put into operation, and you can see what is going on without any other instruments. It is a fast system in that we can see the results of what we are doing almost immediately. Having said that, we can begin work on our encoded motor, but keep in mind this chapter is about closed-loop feedback systems, not encoded motors!

Because all this is fairly complicated for beginners, we will create a series of programs that allow us to creep up on the solution. The relevant programs are developed over the course of this chapter. The programs functions are listed here:

- Hold a motor at its present position; fixed gain to return to position.
- Hold a motor at its present position; fixed and proportional gain to return to position.
- Hold a motor at its present position; fixed, proportional, and integrating gain to return to position.
- Motor position controlled by a potentiometer.
- Motor speed controlled by a potentiometer.
- Motor speed and direction controlled by a potentiometer.
- Motor goes back and forth; potentiometers control gain and distance moved.
- CASE statement used to control gain.
- Determine gain versus speed table for one motor.
- Program segment to determine gain with CASE statement.
- Rudimentary velocity path specification. Slow motion, fast motion, slow motion, stop. Repeat.
- Ramp up and down slowly. Constant rate. Repeat.
- Ramp up, run at constant speed, ramp down, and stop. Repeat.
- Turn a DC motor into an R/C servo.
- Motor speed controlled from an R/C servo signal.

Each program is listed in its entirety because it is very difficult to try to understand a program while looking for program segments here and there at the same time. It is hard enough to keep everything sorted out in your mind as it is.

Discussion

Adding an optical encoder to a motor allows us to determine how fast the motor is spinning, the direction in which it is spinning, and how far it has spun. With this scheme we can control how far the motor moves, how fast it gets to its target location, and what velocity profile it follows on its way to its destination. This essentially is what we define as comprehensive motor control. This is the type of control we need for sophisticated robotic and computer numerically controlled (CNC) machine applications. This is what we need to run a fast, pen-based plotter or, for that matter, any device that needs to coordinate the movement of more than one motor for its competent operation. All multidimensional CNC machines fit this definition (pen plotters, lathes, mills, laser cutters profilers, and many others).

With encoded motors, the first question we are invariably asked is, "How are you going to keep track of the encoder?" (Every one understands that this is a very time-intensive undertaking, so this is an easy question to ask. The hard part is the answer!)

The question is intended to point to the fact that the hard part of running a motor with an attached encoder is reading the encoder constantly and still having time to actually run the motor. It takes up a lot of processor time in a single-processor system. A popular solution is to use a separate chip to read the encoder or to use a processor with built-in hardware to read the encoder. Our solution will be to dedicate one of the cogs to keep track of the encoder and provide us with the current encoder reading whenever we need it. This is equivalent to using a separate chip or processor, only much better because it's all internal to the Propeller chip we are using. We will not need any external wiring to get the job done (other than connecting the encoder up to the Propeller).

The usual scheme used to control the DC motor is called a "PID loop." The letters *P, I,* and *D* represent the three components of the feedback loop that control the motor. A constant, *K,* is needed to take care of friction components. In layman's terms, these items are defined as follows:

- **P** represents the proportional part of the loop.
- **I** represents the integrating function in the loop.
- **D** represents the derivative part of the feedback equation.
- **K** (when used) represents the overall system friction.

For most DC motors, the speed under constant load is proportional to the power sent to the motor. This is the proportional part of the gain equation "P."

If you are running a motor under a variable load, the speed the motor attains will be approximately proportional to the load on the motor. If the load on the motor is increased (and we want to maintain the same speed), we have to add more power to the system to keep the motor speed constant. This is done by adding a little power at a time (again and again in the control loop) until we get to the desired speed (and/or the desired position in time). This is the integrating component of the equation "I."

If we are interested in maintaining the speed of the motor within very tight limits and we want the motor to be where it is expected to be in time, we need to make a calculation about how fast the speed and position of the motor are changing and add an appropriate power component to the motor to keep it within the desired limits. If the motor speed is falling off sharply or its position is falling behind in time, we have to add the power right away instead of integrating it in a little bit at a time, as we were doing with the integrating component. Another way to do this it to calculate where the motor is supposed to be at any one time and make a gain correction based on how great the positional error is. This is how we will do it in our parallel environment. This part of the need for power adjustment is the derivative component of the equation "D."

Because the motor does not start moving until it has overcome the friction in the system, a constant is used to represent the minimum power needed to get it to move. This is the constant "K."

There are many more sophisticated ways to calculate the best way to manage the operation of an encoded DC motor, and they can all work very well. The method selected is usually a function of how fast the system being used to control the motor is. If it is fast enough, and a math package is installed, a complicated mathematical equation can be solved to determine the gain needed. However, in our case we have a small parallel system and only integer math support. Our only recourse is to use a simple scheme like the one described above.

Here is my challenge to you: Go find an easy and comprehensive way to run a DC servo motor anywhere on the Internet or in any book. If easier to undertake than what we will be doing in the following pages, send me the information—*I want to know how it's done.*

The information we get from the motor encoder is also a function of the number of encoder slots in the encoder attached to the motor. As the encoder counts increase, we can determine how far the motor has moved more accurately and we can determine how fast it is moving in a shorter amount of time. However, the time we have to read the encoder between states gets shorter and shorter with increasing encoder counts, and this, too, is critical. If we are using one of the cogs to read the encoder, this is not a critical consideration. However, if we have to use a conventional processor, it is pretty difficult to manage the encoder while running the motor at the same time.

There is a direct relationship between the number of encoder counts and the speed of the motor being controlled because each revolution turned provides the same number of encoder counts. The ideal situation is to have the smallest number of encoder slots that will do the job (but it does not hurt to have larger counts). This has to do with how closely the motor has to be positioned and how closely it has to follow the "time/distance trajectory profile" we are interested in. We have to make changes to the profile often enough to meet the tolerance specification. The motor will tend to depart from the specified motion profile at a variable rate, depending on the changes in the load conditions. The power input correction has to be made often enough to keep the motor position within an acceptable error range for the speed and the distance moved. This is a bit of a handful so we will approach the problem one step at a time.

Although there are exceptions, in almost all cases the motor falls behind what we are trying to have it do. The reason for this is that our control schemes are based on error signals from the motor encoder, and for us to have an error signal, we have to have departed from the desired instantaneous position. As we will see, we use the "time vs. expected position" relationship to try to minimize this lag. The lag is usually in milliseconds and is usually well tolerated by the systems being designed.

When the movements the controlled system makes are rapid, maintaining the exact trajectory becomes more difficult and larger, faster processors and high encoder counts become essential.

Geared systems are used but are frowned upon because they have backlash in them and have to be lubricated and maintained. A direct drive is the drive of choice, and a cog belt–driven system gets the next preferred drive on the selection scale.

DC Servo Motors with Encoders

When we get really serious about running motors with microprocessors, what we are really talking about is running motors that have optical encoders attached to them. This arrangement allows us to control the speed of the motor and to know and control its absolute position. This is what we need to get the really fast changes in speed and position required to build positioning-dependent machines such as plotters, laser cutters, robots, and CNC machines. For the hobbyist, the needs of the robot are more fully met by this arrangement than can be met by any other type of motor-control arrangement inexpensively.

Our interest is in the control of small motors that have relatively coarse encoders attached to them. The encoders provide a two-phase signal, where one phase leads the other by 90 degrees in a 360 degree cycle. The usual signal arrangement is two square waves, as illustrated in Figure 28-2.

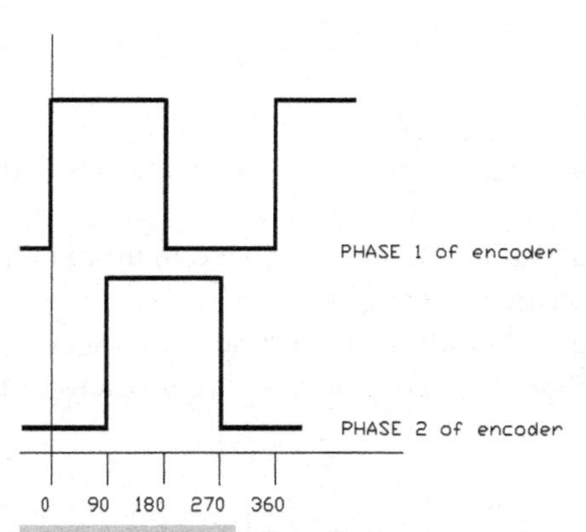

Figure 28-2 Quadrature encoder signals. One signal leads the other by 90 degrees in a 360-degree cycle.

A third channel can provide an indexing pulse during each revolution. This signal can be used to position the motor exactly within any one revolution of motion. Having one repeatable starting position allows all other motor positions to be duplicated exactly. The encoders we are using do not have this third signal. A micro-switch or other device has to be used with the encoder to identify the revolution during which we want to read the index pulse. This means the micro-switch must identify the same revolution every time and the encoder should be set to a position that represents approximately half a revolution of the motor to ensure a repeatable zero position. It's a bit tricky, but you must make sure you understand why this is the case.

Processor Connections

The Propeller chip's ability to address a motor-control process in a parallel environment makes managing the motor control process much easier. We can assign each of the following separable processes within the motor-control process to one of the cogs:

- A cog to read the encoders so we can keep track of where the motor is.
- A cog to manage the LCD so we can see what is going on.
- A cog to read the input potentiometers so we can play with two variables in real time.
- A cog to manage motor power so we can vary the power to the motor as we need to.

Here are the identifications of the Propeller pins we are going to use to connect our motor to the Propeller in these experiments.

The encoder signals will be on P0 and P1:

- **Channel 1** P0
- **Channel 2** P1

The motor amplifier will be connected as follows:

- **Brake** P2
- **PWM** P3
- **Direction** P4

Our amplifier will be one-half of the two-axis Xavien amplifier.

We will read the two potentiometers as follows: Pins P19 to P22 will be connected to the MCP3202 A-to-D converter.

On the converter, we will read line 0 and line 1:

- **Potentiometer 1** line 0
- **Potentiometer 2** line 1

The liquid crystal display will be connected to lines P12 to P18 with a standard 4-bit communications protocol.

We will connect two LEDs to lines 26 and 27 in case we need them to help us look into what is going on in our system at critical junctures.

Not all the lines mentioned will be used for all the experiments. However, this will be our standard setup for playing with an encoded motor, so we are setting it up this way up front.

The Goal

The goal is to tell the motor how far we want it to move and how long we want the move to take—and we want to have the system do it all automatically. The program will calculate everything needed to achieve this goal. How long the ramp up and ramp down will take can be problematic if we want the times calculated automatically, so let's agree that we will accelerate for 25% of the time and decelerate for the same amount of time so we have something concrete to work with. (This is problematic because when you are running more than one motor and each one is following its own velocity path, how and when to speed up or slow down and still stay in sync with the other motors in the system gets complicated. Consider a move where one motor moves 1031 counts and the other moves 13 encoder counts. How should this situation be handled for the coordinated move to be as perfect as possible?)

A discussion of motor control is never complete without a discussion of the PID loop. We will develop some simple programs that help us understand what can be done, but first we need to understand a bit more about the PID loop.

Let's take a closer look at the PID control algorithm definitions. After that, we will write some programs that implement more of the ideas discussed.

PID Control in Greater Detail

Before we go any further, let's get an understanding of what we are talking about when we say that the motor is controlled by a PID loop or equation. The PID loop defines how much energy is to be fed to the motor at any one time during the move. There are four parts to the equation that determines this load. The four parts are referred to as the P, I, D, and K components. The K component is the system friction factor and is sometimes ignored.

- The P part is a proportional component.
- The I part is an integrative component.
- The D part is a derivative component.
- The K part is the overall system friction component

The code controlling the motor power is executed in a loop and the factors are modified each time through the loop as dictated by instantaneous motor performance. If these four components are described properly within the control algorithm, much improved control of the motor will be achieved. Let's look at each component, one at a time, to see what its function is and what it accomplishes. The control scheme we develop does not have to be mathematically perfect to give us good performance. In fact, with Spin and its limited math, a mathematically perfect system cannot be achieved. However, we can get close enough to have a very acceptable operation.

The friction component "K" represents the power that has to be applied to the system to overcome the friction in the system and get the motor turning. Anything less than this, and nothing happens. K can be considered to be a constant though it usually increases with motor speed.

The proportion component "P" tells us that the power to be supplied to the motor is proportional to the speed at which we want to run the motor. This is the largest part of the equation and therefore has to be picked carefully. It will also, of course, depend on the load on (and inertia of) the motor. The faster we want the motor to run, the larger the P component, and the larger the load, the larger the P component. In mathematical terms, the energy provided can be expressed as follows:

$$E = P * X$$

Note *X can be 1 or greater than or less than 1.*

If there are no load changes and the system response is linear (meaning that twice the speed requires twice the power), that is all we need to run the motor. If, however, the load is changing, we have to add to and subtract from the power to keep the motor at the same speed, and we have to keep adding to or subtracting from the power until the motor gets up to proper speed. This is the "I" (integrative) component in the equation. Because it is needed only when there's an error in where the motor should be, I is based on this error. The higher the error, the more we have to add to the power setting to make the motor speed up or slow down to get it to where we need it to be. Needless to say, this is done a bit at a time every time we go through the control loop. In mathematical terms, the energy provided can be expressed as follows:

I = (desired position – actual position) * (a constant or variable of some kind)

Note *The constant is selected to provide as rapid a correction as possible (while keeping the motor under control).*

The derivative component "D" is a measure of the difference in how far the motor moves each time through the cycle with respect to time. (Think about this one for a while.) We are looking to see if the motor is where it is supposed to be with regard to time. To determine this, we need to know where the motor is at any one time, and we need to know where we expect it to be at that time. We therefore need to know precisely (in encoder counts) where the motor is at all times. In mathematical terms, the energy provided can be expressed as follows:

D = (expected position – present position) * (a constant or variable of some kind)

Note *Again, the constant is selected to provide immediate correction.*

One thing this means in simple terms is that there is no need for a change in the power to the motor if the motor is where it needs to be (and/or is moving at the desired speed). We can assign the cogs to give us the desired corrections in real time.

Next, let's develop a few programs to implement and demonstrate the ideas we have just discussed regarding the PID loop. Put a mark on your motor's encoder so you have a way of knowing exactly where the motor is with respect to its home position (see Figure 28-3).

Figure 28-3 **Encoder with an indexing mark on it**

The first program holds the motor at whatever position it is in. If you disturb the motor, it returns it to its starting position. The gain is not programmed; it is controlled by you with potentiometer 1. Play with the gain to see what happens at very low, appropriate, and very high gains when the motor is moved off its home position. Note how long it takes the motor to get back to the zero position as the gain is changed. Note how the motor behaves and at the same time, watch the display. If the gain is too low, it never gets back to its starting position, but there are other interesting things to observe. Can you determine the K factor from this setup?

Now let's proceed with creating the programs listed earlier in the chapter and learn how to control our encoded motor a step at a time.

Holding the Motor Position

The primary control we have to establish over the motor is to place the motor at a specific encoder position and hold that position under all conditions—in other words, hold a motor position once set and have it return to the set position if the motor is disturbed, automatically, no matter what.

If the motor goes over by one or more count in either direction, we move it back to its initial position by providing either positive or negative current to the motor. In order to do this, we have to be able to count the encoder signals as they are generated. We can tell in which direction the motor is moving by determining which of the two phases read from the encoder is leading. When we have established the control we want, the motor will stick at this one position and will return to this position if disturbed. How it returns is determined by the sophistication of the control algorithm, the ideal being a smooth and rapid return to the zero position without overshoot (this is referred to as "perfect damping"). If the gain is too high and too much power is applied, the motor will start to oscillate wildly around the zero position. If not enough power is applied, the motor will not get to the zero position rapidly enough or will not get there at all if the overall system friction is high. In a sophisticated control scheme, the power to return to zero would be a function of how far from the zero position the motor was. It might be changed in real time to reflect the load on the motor and any number of other factors.

What can be done is by and large a function of how fast the processor is and the language being used. A faster processor and Assembly language can be used to do a lot more than the Propeller and the Spin language; however, because we are working with eight processors (a very powerful system indeed), much can be achieved. It will be possible to demonstrate all the techniques used to control encoded DC motors with Spin. What you do to control a motor with Assembly language will pretty much follow what we will do with Spin in the following exercises. The fact that we have eight 32-bit cogs at our disposal makes it possible to do things pretty fast! Parallel processing puts a lot of power at our fingertips, and the power, as we will see, is remarkably easy to use. (All the programming we write will be in Spin, except for a program from the Parallax object exchange to read the encoder.)

The circuitry that we will use to control the encoded DC motor is shown in Figure 28-4.

Note on Encoder Connection *If the motor runs away in a controlled situation, it means that the encoder is connected backward. This can be fixed by reversing the encoder signal leads, reversing the motor leads, or inverting the signals in the software.*

Note *A potentiometer controls the gain in Program 28-1.*

Keeping track of the encoder counts and at the same time managing the motor is not trivial by a long shot. A considerable amount of code has to be processed, and it has to be done constantly. No encoder counts can be dropped. On the other hand, if

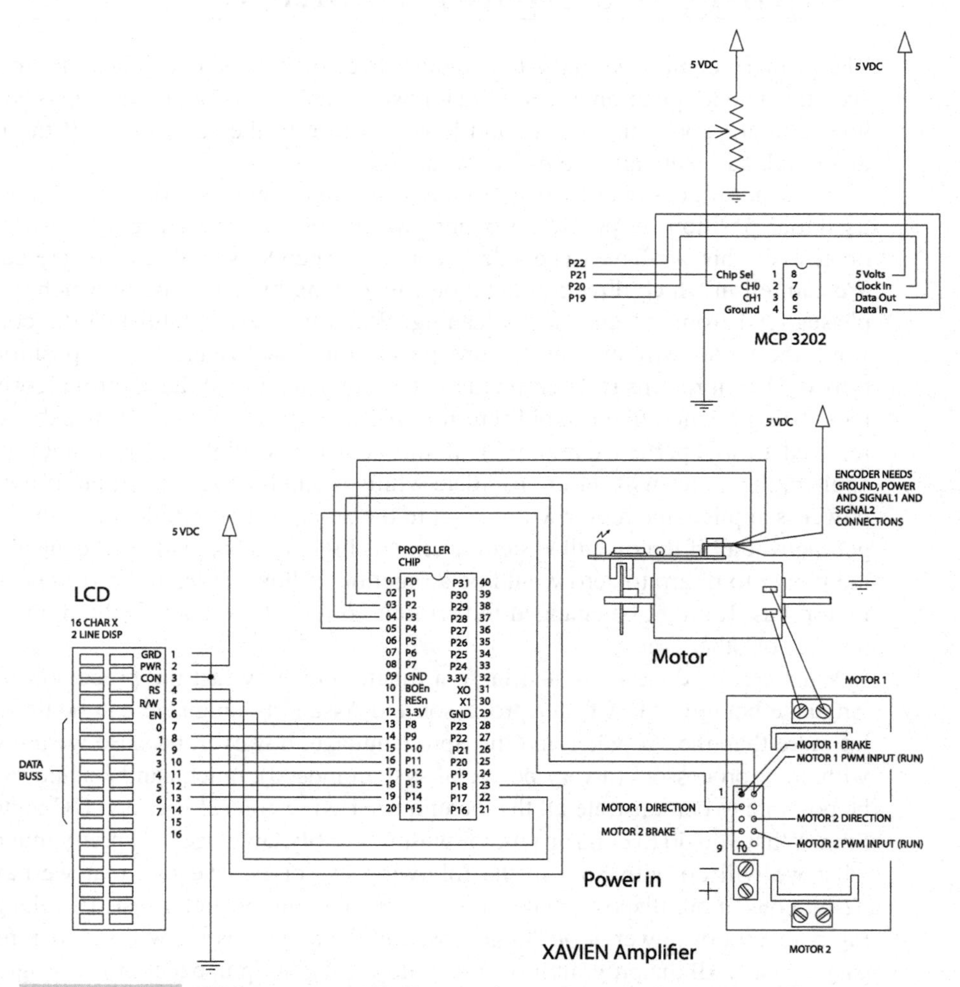

Figure 28-4 Wiring schematic for running a DC motor with an encoder

the encoder tracking can be assigned to one cog, we have a completely different situation. In the eight-processor Propeller system, we can do this with ease.

The pseudocode for holding a motor at one position is as follows:

```
Set an appropriate initial amplifier gain so the motor will move.
Set the current encoder count to zero.
Set the current position as the zero position.
Read the encoder position.
If it has increased, reverse the motor.
If there is no change do nothing.
If it has decreased, move the motor forward.
Go back and read the encoder again.
```

Refinements would consist of schemes for adjusting the power to the motor, based on how far it is from the desired position, in real time. When the motor got close to the desired position, the power supplied would be adjusted to be enough to keep the motor moving at the desired rate. The inertia of the motor and the load on it also play a role in determining the power supplied, meaning that the operation of the system is tuned to one load, unless there is some sort of "tuning-optimization" software in place to compensate for changes in real time. Everything takes processor time, so processor capability is a big factor in running encoded motors. The adjustments we just discussed are implemented in Program 28-1.

Program 28-1 provides the rudimentary control needed to hold a motor on an encoder position, as described previously. Later programs modify this code to add features and refinements.

In Program 28-1, we use five cogs; they are assigned as follows:

- *Cog_0 is the main cog.* This cog starts the other cogs and then goes into a loop that reads the potentiometers. This cog is called "Go."
- *Cog_1 is called Cog_LCD.* It manages the display on the two lines of the LCD. It displays the motor error and the gain (power to) of the motor. The object calls we will use were developed in Chapter 21 (on using the LCD).
- *Cog_2 is called Cog_SetMotPower.* It sets the motor power and direction based on the potentiometer reading and the gain calculations from other cogs. The cogs read the encoder position with the "Encoder" object program. The Encoder program is incorporated into our programs from the Parallax object exchange website. The encoder has to be read in Assembly language and, as beginners, we do not have the expertise to develop this code. On the other hand, in the Propeller system, we all share the code we develop. You have permission to use all the code in this book essentially without restriction or charge, as per the terms of the MIT License (see the preface for details of the license agreement). Therefore, learning how to use the code developed by others is an important part of what we are learning here. The Quadrature Encoder object is the only program we are using from the Propeller Object Exchange (POE). However, we have been using the Utility and LCDRouting4 objects, which we developed ourselves, in our programs on a regular basis.

- *Cog_3 is called Cog_RunMotor.* It is used to set up the counter that manages the PWM needed to run the motor. The PWM is fed to the motor power algorithm in Cog_RunMotor. This is the cog that actually runs the motor by managing the amount of power the motor is fed. The code is based on the code we developed to run a DC motor in Chapter 25.
- *Cog_4 is called Cog_FigGain.* It is used to figure the gain for the motor based on the error signal and the motor position (or whatever the programmer wants). This is the method we will play with to better understand the motor responses.

The code in the following programs is similar, but you'll see some subtle-but-important changes that determine how the readings from the potentiometer are used to control the behavior of the motor and how the power to the motor is determined. Be sure you understand these changes in each program.

Program 28-1 Rudimentary "Holding the Motor on Position" Program

```
{{4 Sep 09        Harprit Sandhu
MotorHoldPos.Spin
Propeller Tool Version 1.2.6
Chapter 28 Programs 01, 02, 03 depending on gain algorithm in use

This program holds an encoded motor at the starting position.
If you disturb the motor it will return to its initial position.
You may have to adjust the gain variable for your particular motor.
You adjust it with POT1. Low values are needed or the motor
oscillates.

Connections are
Amplifier brake       P2
Amplifier PWM         P3
Amplifier direction   P4

Potentiometer read through MCP3202 at P19..P22

LCD on the usual      P12..P18 standard. Set Up.

Revisions
OCT 29 08  Revisions to make variable universal
           added software runaway IF on motor direction
DEC 24 09  Changed to 3202 pots. Uses PotOne

}}
OBJ
  Encoder : "Quadrature Encoder" 'for encoder readings
  LCD     : "LCDRoutines4"       'for the LCD methods
  UTIL    : "Utilities"          'for general methods
```

(continued)

Program 28-1 Rudimentary “Holding the Motor on Position” Program (*continued*)

```
CON
  _CLKMODE=XTAL1 + PLL2X          'The system clock spec
  _XINFREQ     = 5_000_000
  MotorRev     =1   'set this to 0 to reverse a runaway motor
                    'set to 1 for normal operations.
                    'You can also fix this runaway by reversing
                    'the encoder leads S1 and S2.
  BRK       =2      'Connections to the amplifier
  PWM       =BRK+1  'next line
  DIR       =BRK+2  'next line

VAR
  long      Pos[3]          'Create buffer for encoder
  long stack2[25]           'space for Cog_LCD
  long stack3[25]           'space for Cog_SetMotorPower
  long stack4[25]           'space for Cog_RunMotor
  long stack5[25]           'space for Cog_FigureGain
  long Pulswidth            'pulse width
  long PresPosition         'Present Position
  long TargetPosition       'Target Position
  long PositionError        'Positional error
  long MinGain              'Minimum gain
  long Gain                 'Actual gain used
  long MaxGain              'Maxim gain permitted
  long Index                'Counting index
  word Integ                'Integration value
  word PotOne               'Potentiometer reading

PUB Go
  cognew(Cog_LCD,               @stack2)
  cognew(Cog_SetMotPower,       @stack3)
  cognew(Cog_RunMotor(PWM),     @Stack4)
  cognew(Cog_FigGain,           @stack5)
  Encoder.Start(0, 1, 0, @Pos)       'Read the encoder position
  MaxGain:=2000                 'Maxim gain
  MinGain:=100                  'Minimum gain
  repeat
    PotOne:=UTIL.Read3202_0     'Read pot

PRI cog_LCD                          'manage the LCD
  LCD.INITIALIZE_LCD                 'initialize the LCD
  repeat                             'LCD loop
    LCD.POSITION (1,1)               'Go to 1st line 1st space
    LCD.PRINT(STRING("Err=" ))       'Error id
    LCD.PRINT_DEC(PositionError)     'print error
```

(continued)

Program 28-1 Rudimentary "Holding the Motor on Position" Program (*continued*)

```
    LCD.SPACE(2)                         'erase over old data
    LCD.POSITION (2,1)                   'Go to 2nd line 1st space
    LCD.PRINT(STRING("Gain=" ))          'Potentiometer
    LCD.PRINT_DEC(gain)                  'print the gain
    LCD.SPACE(2)                         'erase over old data
    LCD.POSITION (1,11)                  'Go to 1st line 11th space
    LCD.PRINT(STRING("Integ"))           'integrated value
    LCD.SPACE(1)                         'set position for cursor
    LCD.POSITION (2,11)                  'Go to 2nd line 11th space
    LCD.PRINT_DEC(Integ)                 'print the integ
    LCD.SPACE(2)                         'erase over old data

PUB Cog_SetMotPower
  dira[BRK..DIR]~~                       'These pins control the motor amp
  TargetPosition:=0                      'we want to stay where we are
  outa[BRK]~                             'turn off the amp brake
  repeat                                 'loop
    PresPosition:=pos[0]                 'reads the encoder position
    PositionError:=TargetPosition-PresPosition
    case PositionError                   'decision variable
      -1_000_000..-1:                    'range
        if MotorRev                      'negative range
           outa[DIR]~~                   'move positive
        else                             'decide
           outa[DIR]~                    'move negative
        PULSWIDTH:=gain*244/10           'set gain
      -0..0 :                            'range for stopping
         PULSWIDTH:=0                    'gain is set to 0
      1..1_000_000:                      'range
        if MotorRev                      'positive range
           outa[DIR]~                    'move negative
        else                             'decide
           outa[DIR]~~                   'move positive
        PULSWIDTH:=gain*244/10           'set gain

PUB Cog_RunMotor(Pin)|WaveLength,period  ' toggle the output line
  dira[BRK..DIR]~~                     ' the three amplifier lines
  ctra[30..26]:=%00100                 'Set this cog's "A Counter" to run PWM
  ctra[5..0]:=Pin                      'Set "A pin" to Pin#
  frqa:=1                              'Set this counter's frqa value to 1
  PulsWidth:=0                         'Start with pulsewidth=0
  WaveLength:=clkfreq/100              ' time for the pulse width to 10 ms
  period:=cnt                          'Store the current value of the counter
  repeat                               'power PWM routine.
    phsa:=-PulsWidth                   ' high pulse for Pulsewidth counts
```

(*continued*)

Program 28-1 Rudimentary "Holding the Motor on Position" Program (*continued*)

```
    period:=period+WaveLength    'Calculate wave length
    waitcnt(period)              'Wait for the wavelength

PRI Cog_FigGain                  'The Cog we are playing with
  repeat                         'loop
    gain:=PotOne                 'how we define gain here

{{EXERCISE: Play with the equation on the last line to see what happens
to the motor response.

Turn power to motor off and then turn motor by hand and watch what the
LCD shows and look at PWM line P3 to see what the power to the motor
does while you turn POT1.

We will be using this set up repeatedly so leave it all hooked up.
}}
```

The gain is controlled in the three-line PRI Cog_FigGain method. In the next step, let's add some proportionality to this value. Change the PRI Cog_FigGain to what's shown in Program 28-2 and see how the system response changes.

Program 28-2 Segment Adding a Proportional Gain Factor

```
PUB Cog_FigGain                  'power to motor
  repeat                         'loop
    gain:=50+||PositionError     'gain eq goes here. Set at 50 plus
                                 'the absolute value of the error.
```

Program 28-2 holds the motor on position. If you move it off position by turning the motor, it will move back to the set position when you let go of the motor. Note that because the error is added to the fixed gain of the amplifier, the motor turns harder as you get further away from the zero position. However, it is still possible that the motor may not return to the absolute zero position because of the way the fixed portion of the gain (if too low) is set up—there is now a proportional factor but no integration of the gain value in this program. Note that the proportional factor could have a multiplier if we so desired (PotOne is not being used for gain control in this program).

Next, we need to add integration. This is a bit tricky because we need to add it only when the motor is not keeping up with its target position and we need to reset the value to zero once we are at the expected position. Also it is still needed at the end of a move to make sure that the motor gets to its target position. This is because at the end of the move the proportional gain is small and it may not be enough to move the motor and load to its target position without the integration factor.

We can add integration with the code segment shown in Program 28-3. It is very important for you to understand exactly how this algorithm works.

Program 28-3 Segment Adding a Proportional and Integration Factor

```
PRI Cog_FigGain
  repeat                          'loop
    if ||PositionError=<1         'if no error '
       Integ:=0                   'set the integration to zero
       Index:=0                   'set the variable to zero
    Else                          'Otherwise
      index:=index+1              'increment the index counter
    If index==5   ' Max index. The smaller this is the faster the integ
       integ:=integ+1             'increase the integrating function
       index:=0                   'reset the index to 0
    if integ > 2000               'set limit on integrated gain
       integ:=2000                'clamp its value. Max is 4095
    gain:=8+||positionError+integ  'add it all up for the gain.
    gain:=gain <# MaxGain          'gain must be less than MaxGain

{{
EXERCISES: Play with the equations in Cog_FigGain to see what happens
to the motor response as  you change the gain equations.
The value of the maximum index sets how fast the integration proceeds.
The lower the number the faster integ value increases.
The integration max is clamped at 400 to keep you from cutting your
finger on the encoder disk or bending the encoder wheel.
This is a powerful motor so be careful.

Turn power to motor off and then turn motor by hand and watch what the
LCD shows.  Look at PWM line (P3) to see what the power to the motor
does while you turn encoder. Look at it as you change max index.

We will be using this wiring layout for a while so leave it hooked up.

In this program PotOne is read but not used. Divide the 4095 by
a number to get a more realistic value to play with. Use it to change
the variable you want to play with in real time.
}}
```

Be sure to add "Integ" and "index" to the VAR section of the program. The index maximum of 5 determines how fast we integrate. If we make it 1,000, the motor will ramp up much more slowly. You can see all this by turning the power to the motor off and playing with moving the encoder back and forth with your fingers as you watch the LCD. See how the proportional and integrating values change. Do it, but be careful—this motor can ramp up to full power very quickly, and the encoder could cut you! You can make the motor response as stiff as you need for your application. This is a part of the algorithm tuning process for your particular application.

After these additions to the original program, the motor holds its position surprisingly well. You should feel confident that the motor will return to the zero position pretty much under all conditions. Still, we have limited the integration value to 2,000, which could be still higher (4095). It could be stiffer, much stiffer! (Keep in mind that the values selected must never allow the pulse width to exceed 10,000.)

Next, let's modify this program to follow a potentiometer reading, not unlike what an R/C servo does. In order to do this we have to modify the target position so that it is the sum of the original zero position and the potentiometer reading. This is how we build giant servos that follow the count in the target position. This is how the guys on *MythBusters* control their remote-controlled automobiles: one servo for the gas pedal, and one for the steering wheel. As you will see, it's easy when you know how.

Now if we were to add to or subtract from the register that contains the encoder counts to bring the motor to its set position, we would have a rudimentary motor-control algorithm. Of course, many improvements can be made and sophistication added, but basically this is what we are trying to do with all the programs that control encoder-coupled motors.

In Program 28-4, we are adding the count read from a potentiometer to the target position each time through the control loop.

Program 28-4 Motor Position Controlled by a Potentiometer

```
{{4 Sep 09       Harprit Sandhu
MotorHoldPotPos.Spin
Propeller Tool Version 1.2.6
Chapter 28 Program 04

This program moves the encoded motor to a position controlled
by a potentiometer. Range is 4095 encoder counts. This makes a
servo like an R/C hobby servo only much much more powerful.
You may have to adjust the gain variable for your particular motor.

Connections are
Amplifier brake          P2
Amplifier PWM            P3
Amplifier direction      P4

Potentiometer 3202       P19..P22

LCD on the usual   P12..P18

Revisions
Oct 09 08  Revisions to make variables universal
           Added software runaway fin on motor direction
Dec 24 09  Changed over to 4095 pots

}}
OBJ
  Encoder : "Quadrature Encoder"
  LCD     : "LCDRoutines4"          'for the LCD methods
  UTIL    : "Utilities"             'for general methods
```

(continued)

Program 28-4 Motor Position Controlled by a Potentiometer (*continued*)

```
CON
  _CLKMODE=XTAL1 + PLL2X           'The system clock spec
  _XINFREQ    = 5_000_000
  MotorRev    =1                'set this to 0 to reverse a runaway motor
                                'set to 1 for normal operations.
                                'You can also fix this runaway by reversing
                                'encoder leads.
  BRK       =2
  PWM       =BRK+1
  DIR       =BRK+2

VAR
  long     Pos[3]           'Create buffer for encoder
  long stack2[25]           'space for Cog_LCD
  long stack3[25]           'space for Cog_SetMotorPower
  long stack4[25]           'space for Cog_RunMotor
  long stack5[25]           'space for Cog_FigureGain
  long Pulswidth            '
  long PresPosition         '
  long initPosition         '
  long TargetPosition       '
  long PositionError        '
  long MinGain              '
  long Gain                 '
  long MaxGain              '
  long Index                '
  word Integ                '
  word PotOne               '

PUB Go
  cognew(Cog_LCD,              @stack2)
  cognew(Cog_SetMotPower,      @stack3)
  cognew(Cog_RunMotor(PWM),    @Stack4)
  cognew(Cog_FigGain,          @stack5)
  Encoder.Start(0, 1, 0, @Pos)      'Read the encoder position
  MaxGain:=2000
  MinGain:=100
  repeat
      PotOne:=UTIL.Read3202_0

PRI cog_LCD                         'manage the LCD
pulswidth:=89
  LCD.INITIALIZE_LCD                'initialize the LCD
  repeat                            'LCD loop
    LCD.POSITION (1,1)              'Go to 1st line 1st space
    LCD.PRINT(STRING("Err=" ))      'Error
```

(*continued*)

Program 28-4 Motor Position Controlled by a Potentiometer (*continued*)

```
    LCD.PRINT_DEC(PositionError)     'print error
    LCD.SPACE(2)                     'erase over old data
    LCD.POSITION (2,1)               'Go to 2nd line 1st space
    LCD.PRINT(STRING("Gain=" ))      'Potentiometer
    LCD.PRINT_DEC(gain)              'print the pot reading
    LCD.SPACE(2)                     'erase over old data
    LCD.POSITION (1,11)              'Go to 1st line 11th space
    LCD.PRINT(STRING("Integ"))
    LCD.SPACE(1)                     'set position for cursor
    LCD.POSITION (2,11)              'Go to 2nd line 11th space
    LCD.PRINT_DEC(Integ)             'print the index
    LCD.SPACE(2)                     'erase over old data

PUB Cog_SetMotPower
  dira[BRK..DIR]~~                   'These pins control the motor amp
  outa[BRK]~                         'turn off the amp brake
  repeat                             'loop
    PresPosition:=pos[0]             'reads the encoder position
    TargetPosition:=InitPosition+PotOne
    PositionError:=TargetPosition-PresPosition
    case PositionError               'decisions variable
      -1000000..-2:
        if MotorRev                  'negative range
           outa[DIR]~~               'move positive
        else
           outa[DIR]~                'move negative
        PULSWIDTH:=gain*244/10       'set gain
      -1..0 :                        'range for stopping
         PULSWIDTH:=0                'gain is set to 0
      1..1000000:
        if MotorRev                  'positive range
           outa[DIR]~                'move negative
        else
           outa[DIR]~~               'move positive
        PULSWIDTH:=gain*244/10       'set gain

PUB Cog_RunMotor(Pin)|WaveLength,period   ' toggle the output line, set
  dira[BRK..DIR]~~                  ' the three amplifier lines
  ctra[30..26]:=%00100              'Set this cog's "A Counter" to run PWM
  ctra[5..0]:=Pin                   ' "A pin" of cog to Pin# here
  frqa:=1                           'Set this counter's frqa value to 1
  PulsWidth:=0                      'Start with pulsewidth=0
  WaveLength:=clkfreq/100           'Set the pulse width to 10 ms
  period:=cnt                       'Store the current value of the counter
  repeat                            'power PWM routine.
    phsa:=-PulsWidth                'Send high pulse for Pulsewidth count
```

(*continued*)

Program 28-4 Motor Position Controlled by a Potentiometer (*continued*)

```
    period:=period+WaveLength  'Calculate wave length
    waitcnt(period)            'Wait for the wavelength

PRI Cog_FigGain
  repeat                      'loop
    if ||PositionError=<1     'if no error
       Integ:=0               'set the variables to zero
       Index:=0               'set the variables to zero
    Else                      'Otherwise
      index:=index+1          'increment the index counter
    If index==5               'when it reaches set value
       integ:=integ+1         'increase the integrating function
       index:=0               'and reset the index to 0
    if integ > 2000           'set limit on integrated gain
       integ:=2000            'clamp value
    gain:=8+||positionError+integ  'add it all up for the gain.
    gain:=gain <# MaxGain     'gain must be less than MaxGain

{{EXERCISES:
Play with the equation on the last few lines to see what happens
to the motor response to the error signal.
Adjust the line
    TargetPosition:=InitPosition+PotOne
to see what happens
Work on making the response faster. There is a delay right now.
How do you handle the overshoot when the gain has to increase really
fast?

Disconnect power to motor and turn shaft and watch the LCD and the
oscilloscope to see how  the gain changes as the error changes.
This is very instructive and very important to understand well.
}}
```

Next, let's modify the program to move the motor and encoder a few counts at a time, again and again, based on the values we read from a potentiometer. The encoder will be read in the control loop, and the potentiometer reading will be added to the error signal again and again to define the target position. The control algorithm will still be designed to bring the motor to its new zero error position. As we increase the potentiometer signal (that we add to the error signal), the motor will turn faster and faster. To stop it, we have to turn the potentiometer to zero, but this does not return the motor to its original position. The potentiometer now controls the speed of the motor. An error signal (that we can control) is controlling the speed of the motor. The listing for the program that demonstrates this operation is provided in Program 28-5.

The changes to Program 28-4 are as follows:

- Gain made proportional.
- Value read from potentiometer reduced.

Program 28-5 Motor Speed Controlled by a Potentiometer

```
{{4 Sep 09   Harprit Sandhu
HoldMotorPotSpd.Spin
Propeller Tool Version 1.2.6
Chapter 28 Program 05

This program controls the speed of the motor from a potentiometer.
You may have to adjust the gain variable for your particular motor.

Connections are
Amplifier brake          P2
Amplifier PWM            P3
Amplifier direction      P4

Potentiometer           P19

LCD on the usual    P12..P18

Revisions
Oct 09 08  Revisions to make variables universal
           Added software runaway fin on motor direction
Dec 24 09  Changed over to 4095 pots

}}
OBJ
  Encoder : "Quadrature Encoder"
  LCD     : "LCDRoutines4"        'for the LCD methods
  UTIL    : "Utilities"           'for general methods

CON
  _CLKMODE=XTAL1 + PLL2X          'The system clock spec
  _XINFREQ    = 5_000_000
  MotorRev    =1                 'set this to 0 to reverse a runaway motor
                                 'set to 1 for normal operations.
                                 'You can also fix this runaway by reversing
                                 'encoder leads.
  BRK       =2
  PWM       =BRK+1
  DIR       =BRK+2

VAR
  long     Pos[3]            'Create buffer for encoder
  long stack2[25]            'space for Cog_LCD
  long stack3[25]            'space for Cog_SetMotorPower
  long stack4[25]            'space for Cog_RunMotor
  long stack5[25]            'space for Cog_FigureGain
  long Pulswidth
```

(continued)

Program 28-5 Motor Speed Controlled by a Potentiometer (*continued*)

```
  long PresPosition
  long initPosition
  long TargetPosition
  long PositionError
  long MinGain
  long Gain
  long MaxGain
  long Index
  word Integ
  word PotOne

PUB Go
  cognew(Cog_LCD,             @stack2)
  cognew(Cog_SetMotPower,     @stack3)
  cognew(Cog_RunMotor(PWM),   @Stack4)
  cognew(Cog_FigGain,         @stack5)
  Encoder.Start(0, 1, 0, @Pos)     'Read the encoder position
  MaxGain:=2000
  MinGain:=100
  TargetPosition:=0                'we want to stay where we are
  repeat
      PotOne:=UTIL.Read3202_0

PRI cog_LCD                        'manage the LCD
pulswidth:=89
  LCD.INITIALIZE_LCD               'initialize the LCD
  repeat                           'LCD loop
    LCD.POSITION (1,1)             'Go to 1st line 1st space
    LCD.PRINT(STRING("Err=" ))     'Error
    LCD.PRINT_DEC(PositionError)   'print error
    LCD.SPACE(2)                   'erase over old data
    LCD.POSITION (2,1)             'Go to 2nd line 1st space
    LCD.PRINT(STRING("Gain=" ))    'Potentiometer
    LCD.PRINT_DEC(gain)            'print the pot reading
    LCD.SPACE(2)                   'erase over old data
    LCD.POSITION (1,11)            'Go to 1st line 11th space
    LCD.PRINT(STRING("Integ"))
    LCD.SPACE(1)                   'set position for cursor
    LCD.POSITION (2,11)            'Go to 2nd line 11th space
    LCD.PRINT_DEC(Integ)           'print the index
    LCD.SPACE(2)                   'erase over old data

PUB Cog_SetMotPower
  dira[BRK..DIR]~~                 'These pins control the motor amp
  outa[BRK]~                       'turn off the amp brake
```

(*continued*)

Program 28-5 Motor Speed Controlled by a Potentiometer (*continued*)

```
repeat                              'loop
  PresPosition:=pos[0]              'reads the encoder position
  TargetPosition:=TargetPosition+PotOne/256/3
  PositionError:=TargetPosition-PresPosition
  case PositionError                'decision variable
    -1_000_000..-1:
      if MotorRev                   'negative range
         outa[DIR]~~                'move positive
      else
         outa[DIR]~                 'move negative
      PULSWIDTH:=gain*244/10        'set gain
    -0..0 :                         'range for stopping
       PULSWIDTH:=0                 'gain is set to 0
    1..1_000_000:
      if MotorRev                   'positive range
         outa[DIR]~                 'move negative
      else
         outa[DIR]~~                'move positive
      PULSWIDTH:=gain*244/10        'set gain


PUB Cog_RunMotor(Pin)|Cycle_time,period  ' toggle the output line, set
  dira[BRK..DIR]~~                 ' the three amplifier lines
  ctra[30..26]:=%00100             'Set this cog's "A Counter" to run PWM
  ctra[5..0]:=Pin                  'Set "A pin" to Pin#
  frqa:=1                          'Set this counter's frqa value to 1
  PulsWidth:=0                     'Start with pulsewidth=0
  Cycle_time:=clkfreq/100          ' time for the pulse width to 10 ms
  period:=cnt                      'Store the current value of the counter
  repeat                           'power PWM routine.
    phsa:=-PulsWidth               ' high pulse for Pulsewidth counts
    period:=period+Cycle_time      'Calculate cycle time
    waitcnt(period)                'Wait for the cycle time

PRI Cog_FigGain
  repeat                        'loop
    if ||PositionError=<1       'if no error '
       Integ:=0                 'set the variables to zero
       Index:=0                 'set the variables to zero
    Else                        'Otherwise
      index:=index+1            'increment the index counter
    If index==5                 'when it reaches set value
       integ:=integ+1           'increase the integrating function
       index:=0                 'and reset the index to 0
    if integ > 2000             'set limit on integrated gain
       integ:=2000              'clamp value
```

(continued)

Program 28-5 Motor Speed Controlled by a Potentiometer (*continued*)

```
    gain:=8+||positionError+integ/4   'add it all up for the gain.
    gain:=gain <# MaxGain    'gain must be less than MaxGain

{{EXERCISES:
Play with the equation on the last few lines to see what happens
to the motor response to the error signal.
Adjust the
    TargetPosition:=TargetPosition+PotOne
line to see what happens.
Work on making the response faster. There is a delay right now.
How do you handle the overshoot when the gain has to increase real
fast?

Disconnect power to motor and turn shaft and watch the LCD and the
oscilloscope to see how  the gain changes as the error changes.

Connect the oscilloscope to the PWM line.
This is very instructive and very important to understand well.
}}
```

In Program 28-5, the potentiometer position is added to the motor position register and the program tries to bring the motor position/error back to the new zero. As you turn the potentiometer further and further, the motor moves faster and faster because the error is increasing more and more rapidly. The motor keeps moving because the potentiometer position is added to the error register each time through the loop. Watch the error count in the display as you play with the potentiometer. It is not zero and reflects the motor error and the gain. *The error determines the gain and therefore the speed!* If there is no error, we don't need to do anything, so everything depends on the error. The goal is to determine the error and find ways to keep it minimized at all times. It is imperative that you understand exactly how cog_FigGain works in every detail in this program. This is the most sophisticated gain algorithm in the book even though it is not used in some of the later programs. It can be incorporated in them if you like.

What we see in Program 28-5 is a simple motor speed controller operated from a potentiometer. This program fails if the target register overflows and does not tolerate high motor gains. What needs to be done to fix this? (We have a huge target register, but eventually this will be a problem that needs to be addressed. The gain is defined as 12 bits, so it must not be allowed to exceed 4,095. However, that is too high, so it has been reduced to keep things manageable.)

As currently written, Program 28-5 moves the motor in one direction only. We need the potentiometer to move the motor back and forth, and we also need it to control the motor's speed. If we subtract 2,048 from the potentiometer reading, we can interpret the potentiometer reading as a value between –2,048 and +2,047, with 0 being the "no movement" position. We will rewrite the code so the motor will respond accordingly.

After we incorporate the code into the program, we get what is shown in Program 28-6. We have added the code we need to interpret the potentiometer reading as needed for motor reversal.

Program 28-6 Control the Motor Speed and Direction from a Potentiometer

```
{{14 Sep 09        Harprit Sandhu
MotorHoldPotSpdDir.Spin
Propeller Tool Version 1.2.6
Chapter 28  Program 06

This program runs the motor back or forth at a speed and direction
as controlled by the potentiometer.

You may have to adjust the gain variable for your particular motor.

Connections are
Amplifier brake       P2
Amplifier PWM         P3
Amplifier direction   P4

Potentiometer         P19
LCD on the usual P12..P18
Revisions

}}
OBJ
  Encoder : "Quadrature Encoder"
  LCD     : "LCDRoutines4" 'for the LCD methods
  UTIL    : "Utilities"  'for general methods

VAR
  long     Pos[3]      'Create buffer for two encoders plus room
                       ' for delta position support of 1st encoder)
  long stack2[35]      'space for Cog
  long stack3[35]      'space for Cog
  long stack4[35]      'space for Cog
  word pcount          '
  long pulswidth       '
  word dcount

CON
  _CLKMODE=XTAL1+ PLL2X    'The system clock spec
  _XINFREQ = 5_000_000
  PotMax   = 100          'The maximum power to the motor OUT OF 255
  PotMin   = 0            'The minimum power to the motor OUT OF 255
  AmpBrk   = 2
  AmpPWM   = AmpBrk+1
  AmpDir   = AmpBrk+2
```

(continued)

Program 28-6 Control the Motor Speed and Direction from a Potentiometer (*continued*)

```
PUB Go
  cognew(cog_LCD, @stack2)          'start a new cog
  cognew(SetMotorPower, @stack3)    'start a new cog
  cognew(RunMotor(AmpPWM),@Stack4)       'start a new cog
  Encoder.Start(0, 1, 0, @Pos)      'read encoder
  dira[AmpBrk]~~                    'make output
  outa[AmpBrk]~                     'turn brake off

PUB SetMotorPower
dira[AmpDir]~~
pos[0]:=0
  repeat
    pcount:=UTIL.Read3202_0/16  'get the pot reading from the utilities
    case pcount                 'depends on pcount
      0..127:                   'check for center position
        outa[AmpDir]~           'forward direction set
        pcount:=(127-pcount)*2  ' double the 127 to get near 254
      128..129:                'zero position
        pcount:=0              'clamp to 0
      130..255:                'or
        outa[AmpDir]~~         'reverse direction
        pcount:=(pcount-128)*2  ' double the 127 to get near 254
    if pcount<PotMin           'check Min value
       pcount:=Potmin          'Clamp to Min value
    if pcount>PotMax           'check Max value
        pcount:=Potmax         'clamp to Max value
    PulsWidth:=pcount*450      'mult reading to get count needed
    dcount:=pcount

PRI cog_LCD
  LCD.INITIALIZE_LCD                'initialize the LCD
  repeat                            'LCD loop
    LCD.POSITION (1,1)              'Go to 1st line 1st space
    LCD.PRINT(STRING("EncCnt=" ))   'Potentiometer position ID
    LCD.PRINT_DEC(Pos[0])           'print the pot reading
    LCD.SPACE(3)                    'erase over old data
    LCD.POSITION (2,1)              'Go to 2nd line 1st space
    LCD.PRINT(STRING("Pot    =" ))  'Potentiometer position ID
    LCD.PRINT_DEC(dcount)           'print the pot reading
    LCD.SPACE(3)                    'erase over old data

PUB RunMotor(Pin)|WaveLength,period    ' toggle the output line, set
  dira[2..4]~~                      ' the three amplifier lines
  ctra[30..26]:=%00100              'Set this cog's "A Counter" to run PWM
  ctra[5..0]:=Pin                   'Set the "A pin" of this cog to Pin
  frqa:=1                           'Set this counter's frqa value to 1
```

(continued)

Program 28-6 Control the Motor Speed and Direction from a Potentiometer (*continued*)

```
PulsWidth:=0                      'Start with position=0
WaveLength:=clkfreq/100           'Set pulse width to 10 ms
period:=cnt                       'Store the current value of the counter
repeat                            'power PWM routine.
  phsa:=-PulsWidth                ' high pulse for Pulsewidth counts
  period:=period+WaveLength       'Calculate wave length
  waitcnt(period)                 'Wait for the wavelength
```

In Program 28-6, the code has been modified so that the gain and the direction of motion are a function of the potentiometer reading. It is integral to the program—there is no gain-controlling cog. The constants PotMin and PotMax have also been added to the code but are set to minimum (0) and maximum (100) values. As you play with the program, set these to various values to see how the program responds to the changes you make. Watch the potentiometer position on the LCD. Make it go to 0, the mid position to stop the motor.

Next, we need a program that moves the motor back and forth a few counts at a time automatically. This will allow us to play with the gain in the system in real time without having to worry about actually running the motor back and forth. In this program we add a known, fixed count (read from a potentiometer) to the target position and then move to that position. We must add code (the StrtFlag method) to allow the program to wait until each move is completed. The system is not aware of the motor motion and therefore does not do this automatically. Then we pause for a short time, subtract the same count from the target position, and move back to where we started from. We then pause again and repeat the program. The code that does just that is provided in Program 28-7. Note that the only change between the positive and the negative moves is the addition or subtraction from the target position. It is important to understand that we are always trying to get to the target position. In later programs we will create "time-dependent target position profiles" to create the move profiles we need.

Note *We will play with similar programs as we develop our control algorithms.*

Program 28-7 Motor Moves Back and Forth. Motor Gain and Distance Controlled by Potentiometers

```
{{04 Nov 09    Harprit Sandhu
MotorBckForPotGain.Spin
Propeller Tool Version 1.2.6
Chapter 28  Program 07

This program runs the motor back and forth while you vary
the gain and distance moved with two potentiometers.
Pot1 controls the distance and
Pot2 controls the gain.
```

(continued)

Program 28-7 Motor Moves Back and Forth. Motor Gain and Distance Controlled by Potentiometers (*continued*)

```
How we control the gain is determined by the method that
sets the gain at the bottom of the programs.

We will use the information gained here to design the equations
to set the gain to make the move as fast as possible without
overshoot.

You may have to adjust the parameters for your particular motor.

Connections are
Amplifier brake       P2
Amplifier PWM         P3
Amplifier direction   P4

Potentiometer        P19

LCD on the usual P8..P18

}}
OBJ
  Encoder : "Quadrature Encoder"
  LCD     : "LCDRoutines4" 'for the LCD methods
  UTIL    : "Utilities"  'for general methods

CON
  _CLKMODE=XTAL1+ PLL2X            'The system clock spec
  _XINFREQ = 5_000_000

VAR
  long     Pos[3]          'Create buffer for encoder
  long stack2[25]          'space for Cog_LCD
  long stack3[25]          'space for Cog_SetMotorPower
  long stack4[25]          'space for Cog_RunMotor
  long stack5[25]          'space for Cog_FigureGain
  long stack6[25]          'space for Cog_Start
  long stack7[25]          'space for readpots
  long pulswidth           '
  long PresentPosition     '
  long TargetPosition      '
  long PositionError       '
  long gain                '
  long dgain               'displayed gain
  word PotReading          '
  word startFlag           '
  long startPosition       '
```

(*continued*)

Program 28-7 Motor Moves Back and Forth. Motor Gain and Distance Controlled by Potentiometers (*continued*)

```
  long distance            '
  long pot1                '
  long Pot2                '

PUB Go
  cognew(cog_LCD, @stack2)
  cognew(SetMotorPower, @stack3)
  cognew(RunMotor(3),@Stack4)
  cognew(figureGain,  @stack5)
  cognew(StrtFlag,  @stack6)
  cognew(ReadPots,  @stack7)
  Encoder.Start(0, 1, 0, @Pos)  'Reads the encoder position
  startPosition:=Pos[0]         'position read
  distance:=42*0                'number of 1/4th revs is 8 for 1 rev
  repeat
    targetPosition:=StartPosition+distance
    waitcnt(24_000+cnt)
    repeat while startFlag==0
    {The start flag is needed to make sure that one instruction
    gets done before the next instructions starts. This is
    managed with the start flag. Try eliminating these
    instructions to see what happens. Is there a better way to
    manage this? As long as the start flag is 0 the program holds
    at this line. It took me a while to get this one figured out!
    }
    waitcnt(2_000_000+cnt)
    targetPosition:=startPosition
    waitcnt(24_000+cnt)
    repeat while startFlag==0
    waitcnt(2_000_000+cnt)

PRI SetMotorPower
  dira[2..4]~~                      'These pins control the motor amp
  repeat                            'loop
    PresentPosition:=pos[0]         'read the encoder position
    PositionError:=TargetPosition-PresentPosition
    case PositionError              'decision variable
      -100_000_000..-2:             'negative range
         outa[4]~~                  'move in position direction
         PULSWIDTH:=gain*450        'set gain
      -1..1 :                       'range for stopping
         PULSWIDTH:=0*450           'gain is 0
      2..100_000_000:               'positive range
         OUTA[4]~                   'move in negative direction
         PULSWIDTH:=gain*450        'set gain
```

(*continued*)

Program 28-7 Motor Moves Back and Forth. Motor Gain and Distance Controlled by Potentiometers (*continued*)

```
PRI cog_LCD                                 'manage the LCD
  LCD.INITIALIZE_LCD                        'initialize the LCD
  repeat                                    'LCD loop
    LCD.POSITION (1,1)                      'Go to 1st line 1st space
    LCD.PRINT(STRING("Er=" ))               'Error
    LCD.PRINT_DEC(PositionError)            'print error
    LCD.SPACE(2)                            'erase over old data
    LCD.POSITION (1,10)                     'Go to 1st line 10th space
    LCD.PRINT_DEC(pot1)                     'print error
    LCD.SPACE(2)                            'erase over old data
    LCD.POSITION (1,14)                     'Go to 1st line 14th space
    LCD.PRINT_DEC(pot2)                     'erase over old data
    LCD.SPACE(2)                            'erase over old data
    LCD.POSITION (2,1)                      'Go to 2nd line 1st space
    LCD.PRINT(STRING("Gain=" ))             'Potentiometer
    LCD.PRINT_DEC(dgain)                    'print the pot reading
    LCD.SPACE(2)                            'erase over old data
    LCD.POSITION (2,10)                     'Go to 2nd line 10th space
    LCD.PRINT(STRING("Start=" ))            'Potentiometer
    LCD.PRINT_DEC(StartFlag)                'print the pot reading
    distance:=pot1*21


PRI RunMotor(Pin)|WaveLength,period        ' toggle the output line, set
  dira[2..4]~~                             ' the three amplifier lines
  ctra[30..26]:=%00100                   'Set this cog's "A Counter" to run PWM
  ctra[5..0]:=Pin                        'Set the "A pin" of this cog to Pin=3
  frqa:=1                                'Set this counter's frqa value to 1
  PulsWidth:=0                           'Start with position=0
  WaveLength:=clkfreq/100                'Set the time for the wave L to 10 ms
  period:=cnt                            'Store the current value of the counter
  repeat                                 'power PWM routine.
    phsa:=-PulsWidth                     ' high for Pulsewidth counts
    period:=period+WaveLength            'Calculate wave length
    waitcnt(period)                      'Wait for end of wavelength

PRI StrtFlag                            'used to determine if motor completed move
  repeat
    if ||PositionError<2                'is move almost done
      startFlag:=1                      'if it is it's ok to start
    else                                '1f not
      startFlag:=0                      'do not start next move
```

(*continued*)

Program 28-7 Motor Moves Back and Forth. Motor Gain and Distance Controlled by Potentiometers (*continued*)

```
PUB figureGain                      'power to motor
  repeat                            'loop
    case ||PositionError            'pot is NOT READ
      0..20:gain:=pot2/10           'in these comparisons
      21..80:gain:=pot2/4           'the gain is set by the
      81..200:gain:=pot2/2          'amount of the error
      101..80000:gain:=pot2         'in the position.
    gain#>=8
    dgain:=gain

PRI Readpots
  repeat
    pot1:=UTIL.Read3202_0/16
    pot2:=UTIL.Read3202_1/16
```

Play with this program by changing the positions of the potentiometers. Make other changes to various entities to see how the system responds. This will give you important clues about how you might proceed with motor control.

As you play with the constants and variables in this program, you will notice that essentially the gain on the motor needs better control (we lose control at high and low gains and when making short moves). In the next steps, we will improve on the operation of the motor as affected by the error in the move position (and thereby the gain on the amplifier).

The encoder I used has 42 full quadrature cycles per revolution, and the program reads every encoder change on each of the two channels, to give 168 counts per revolution. The average move we make needs to consist of about 25 revolutions of the motor. (Approximately 4,000 counts in round numbers, in each direction, as we play with the system.) This move will be long enough to allow us to develop the control we need, yet short enough not to waste time. There will be time enough for the motor to speed up, run at a constant speed, and then slow down and stop at the desired point. Each time it stops, the encoder must be in the same position if things are right because we moved forward and backward the same amount.

We can use a CASE statement to set the motor gain depending on the error signal. Here we are turning over the calculation of the motor gain to a computerized function that we can base on whatever we want. We can now respond to the error signal in any way we want as long as we can describe our needs with a CASE statement. Of course, we do not have to use a CASE statement, but in Spin this is the easiest way to achieve our goal at our current level of expertise.

We implement these ideas in Program 28-8. There is no potentiometer input in this program, although that could be added. Here, we control the system with the gain equations to see what happens. You have to change the program and then download it for each iteration to see what the effect of the changes you made to the gain equation is.

Program 28-8 Motor Moves Back and Forth. Motor Speed/Gain Controlled by CASE Statement, Move Distance Preset.

```
{{14 Sep 09   Harprit Sandhu
MotorBakForPlayGain.Spin
Propeller Tool Version 1.2.6
Chapter 28  Program 08

This program runs the motor back and forth a fixed number
of counts while you can vary the gain with the equations
at the bottom of the programs to see what happens to
the motor dampening function.  We will use the information
gained here to design the equations to set the gain to make
the move as fast as possible without overshoot.

You may have to adjust the parameters for your particular motor.

Connections are
Amplifier brake        P2
Amplifier PWM          P3
Amplifier direction    P4
Potentiometer         P19
LCD on the usual P8..P18
}}
OBJ
  Encoder : "Quadrature Encoder"
  LCD     : "LCDRoutines4" 'for the LCD methods
  UTIL    : "Utilities"   'for general methods

CON
  _CLKMODE=XTAL1+ PLL2X   'The system clock spec
  _XINFREQ = 5_000_000
  distance =168*8         'full revs

VAR
  long     Pos[3]         'Create buffer for encoder
  long stack2[25]         'space for Cog_LCD
  long stack3[25]         'space for Cog_SetMotorPower
  long stack4[25]         'space for Cog_RunMotor
  long stack5[25]         'space for Cog_FigureGain
  long stack6[25]         'space for Cog_Start
  long pulswidth          '
  long PresentPosition    '
  long gain               '
  long TargetPosition     '
  long PositionError      '
  byte startFlag          '
  long startPosition      '
```

(continued)

Program 28-8 Motor Moves Back and Forth. Motor Speed/Gain Controlled by CASE Statement, Move Distance Preset. (*continued*)

```
PUB Go
  cognew(cog_LCD, @stack2)
  cognew(SetMotorPower, @stack3)
  cognew(RunMotor(3),@Stack4)
  cognew(figureGain,  @stack5)
  cognew(StaartFlag,  @stack6)
  Encoder.Start(0, 1, 0, @Pos)       'Reads the encoder position
  startPosition:=Pos[0]
  repeat
    targetPosition:=StartPosition+distance
    waitcnt(10000+cnt)
    repeat while startFlag==0
    waitcnt(2_000_000+cnt)
    targetPosition:=startPosition
    waitcnt(10000+cnt)
    repeat while startFlag==0
    waitcnt(2_000_000+cnt)

PUB SetMotorPower
  dira[2..4]~~                        'These pins control the motor amp
  repeat                              'loop
    PresentPosition:=pos[0]           'read the encoder position
    PositionError:=TargetPosition-PresentPosition
    case PositionError                'decision variable
      -100_000_000..-2:               'negative range
         outa[4]~~                    'move in position direction
         PULSWIDTH:=gain*450          'set gain
      -1..1 :                         'range for stopping
         PULSWIDTH:=0*450             'gain is 0
      2..100_000_000:                 'positive range
         OUTA[4]~                     'move in negative direction
         PULSWIDTH:=gain*450          'set gain

PRI cog_LCD                           'manage the LCD
  LCD.INITIALIZE_LCD                  'initialize the LCD
  repeat                              'LCD loop
    LCD.POSITION (1,1)                'Go to 1st line 1st space
    LCD.PRINT(STRING("Err=" ))        'Error
    LCD.PRINT_DEC(PositionError)      'print error
    LCD.SPACE(5)                      'erase over old data
    LCD.POSITION (2,1)                'Go to 2nd line 1st space
    LCD.PRINT(STRING("Gain=" ))       'gain
    LCD.PRINT_DEC(gain)               'print the pot reading
    LCD.SPACE(2)                      'erase over old data
```

(continued)

Program 28-8 Motor Moves Back and Forth. Motor Speed/Gain Controlled by CASE Statement, Move Distance Preset. (*continued*)

```
    LCD.POSITION (2,10)               'Go to 2nd line 1st space
    LCD.PRINT(STRING("Start=" ))      'Flag
    LCD.PRINT_DEC(StartFlag)          'print flag status

PUB RunMotor(Pin)|WaveLength,period   ' toggle the output line, set
  dira[2..4]~~                        'g the three amplifier lines
  ctra[30..26]:=%00100               'Set this cog's "A Counter" to run PWM
  ctra[5..0]:=Pin                    'Set the "A pin" of this cog to Pin
  frqa:=1                            'Set this counter's frqa value to 1
  PulsWidth:=0                       'Start with position=0
  WaveLength:=clkfreq/100            'Set pulse width to 10 ms
  period:=cnt                        'Store the current value of the counter
  repeat                             'power PWM routine.
    phsa:=-PulsWidth                 ' high pulse for Pulsewidth counts
    period:=period+WaveLength        'Calculate wave length
    waitcnt(period)                  'Wait for the wavelength

PUB figureGain                       'power to motor 0-255
  repeat                             'loop
    case ||PositionError             'pot is NOT READ
      1..20:gain:=6                  'in these comparisons
      21..80:gain:=10                'the gain is set by the
      81..200:gain:=40               'amount of the error
      101..80000:gain:=80            'in the position.

PRI StaartFlag       'The start flag is used to inhibit the execution of
  repeat             'the next move to make sure that each move executes
    if ||PositionError<2   'fully. Then wait comd lets you see that the
      startFlag:=1      'motor stopped before proceeding to reversing
    else                  'the motor.  See main Cog above for more.
      startFlag:=0
```

After playing with the CASE constructs at the end of Program 28-8 in the figureGain method for a while, you will start to get a feel for how to design a gain algorithm that will give you a satisfactory solution for the motor/load combination you are working with. If the language we are using had the ability to solve equations and is fast enough, we could, of course, use that facility.

Let's write a program that gives us some idea of how the motor responds to the power supplied to it. For each gain, we want to know how fast the motor moves as expressed in encoder counts per second. From the data, we will create a table.

As we have done previously, in Program 28-9 we are controlling the motor speed with the potentiometer, but in this case we want to display the speed and potentiometer setting, so we can create a "power vs. speed" table (see Table 28-1).

Program 28-9 Creating a Motor Speed vs. Motor Power Table. Gain for Each Reading Controlled Manually with a Potentiometer.

```
{{14 Sep 09   Harprit Sandhu
MotorPotSpeedTable.Spin
Propeller Tool Version 1.2.6
Chapter 28 Program 09

This program controls speed of motor from the pot.
Displays speed and pot setting.
You may have to adjust the gain variable for your particular motor.

Connections are
Encoder 1             P0
Encoder 2             P1
Encoder 3             5 Volts
Encoder 4             Ground
  Also need power and ground for encoder connected
Amplifier brake       P2    Xavien  1
Amplifier PWM         P3    Xavien  2
Amplifier direction   P4    Xavien  3

Potentiometer        P19

LCD on the usual P8..P18
The values read are based on a 13.5 volts power
supply for the motor. (Table in book)

}}
OBJ
  Encoder : "Quadrature Encoder"
  LCD     : "LCDRoutines4" 'for the LCD methods
  UTIL    : "Utilities"    'for general methods

CON
  _CLKMODE=XTAL1+ PLL2X         'The system clock spec
  _XINFREQ = 5_000_000

VAR
  long     Pos[3]          'Create buffer for encoder
  long stack2[25]          'space for Cog_LCD
  long stack3[25]          'space for Cog_SetMotorPower
  long stack4[25]          'space for Cog_RunMotor
  long stack5[25]          'space for Cog_FigureGain
  long pulswidth           '
  long speed               '
```

(continued)

Program 28-9 Creating a Motor Speed vs. Motor Power Table. Gain for Each Reading Controlled Manually with a Potentiometer. (*continued*)

```
  long start                       '
  long end                         '

PUB Go
  cognew(cog_LCD, @stack2)
  cognew(SetMotorPower, @stack3)
  cognew(RunMotor(3),@Stack4)
  cognew(FigureCounts, @stack5)
  Encoder.Start(0, 1, 0, @Pos)        'Read the encoder position

PUB SetMotorPower
  dira[2..4]~~                        'These pins control the motor amp
  repeat                              'loop
    PulsWidth:=Util.Read3202_0/16     'read the 4095 pot position to a byte

PRI cog_LCD                          'manage the LCD
  LCD.INITIALIZE_LCD                 'initialize the LCD
  repeat                             'LCD loop
    LCD.POSITION (1,1)               'Go to 1st line 1st space
    LCD.PRINT(STRING("Speed=" ))     'Error
    LCD.PRINT_DEC(Speed)             'print error
    LCD.SPACE(3)                     'erase over old data
    LCD.POSITION (2,1)               'Go to 2nd line 1st space
    LCD.PRINT(STRING("Gain =" ))     'Potentiometer
    LCD.PRINT_DEC(pulswidth)              'print the pot reading
    LCD.SPACE(3)                     'erase over old data

PUB RunMotor(Pin)|Cycle_time,period    ' toggle the output line, set
  dira[2..4]~~                        ' the three amplifier lines
  ctra[30..26]:=%00100             'Set this cog's "A Counter" to run PWM
  ctra[5..0]:=Pin                  'Set the "A pin" of this cog to Pin
  frqa:=1                          'Set this counter's frqa value to 1
  PulsWidth:=0                     'Start with position=0
  Cycle_time:=clkfreq/100          'Set the pulse width to 10 ms
  period:=cnt                      'Store the current value of the counter
  repeat                           'power PWM routine.
    phsa:=-PulsWidth*24*16         ' high pulse for Pulsewidth counts
    period:=period+Cycle_time      'Calculate cycle time
    waitcnt(period)                'Wait for the cycle time

PUB FigureCounts                   'encoder counts
  repeat                           'loop
    start:=Pos[0]                  'read encoder
    waitcnt(clkfreq+cnt)           'pause 1 sec
    end:=Pos[0]                    'read encoder
    speed:=(end-start)             'counts moved
```

Note *Speed is in encoder counts per second; 42-slot encoder, four counts per slot. Supply voltage is 13.5 volts; gain is in parts of the 255 full gain. (The results for your motor will not be the same as mine.)*

With the information in Table 28-1, we can define the friction components and decide on the approximate proportional gain factors as we develop more sophisticated control algorithms.

TABLE 28-1 POWER/GAIN VS. MOTOR SPEED IN ENCODER COUNTS/SECOND

GAIN	SPEED	GAIN	SPEED
	0	40	2720
2	11	45	3055
3	23	50	3386
4	85	55	3660
5	142	70	4700
6	152	80	5200
7	320	90	6100
8	400	100	6800
9	460	110	7400
10	560	120	8200
11	640	130	8800
12	710	140	9800
13	800	150	10700
14	850	160	11700
15	940	170	12500
16	1010	180	13700
17	1080	190	14600
18	1160	200	15600
19	1200	210	16700
20	1300	220	17900
21	1370	230	18800
22	1450	240	19800
23	1540	250	21000
24	1630	230	21100
25	1660	240	21200
30	2000	250	21300
35	2400		

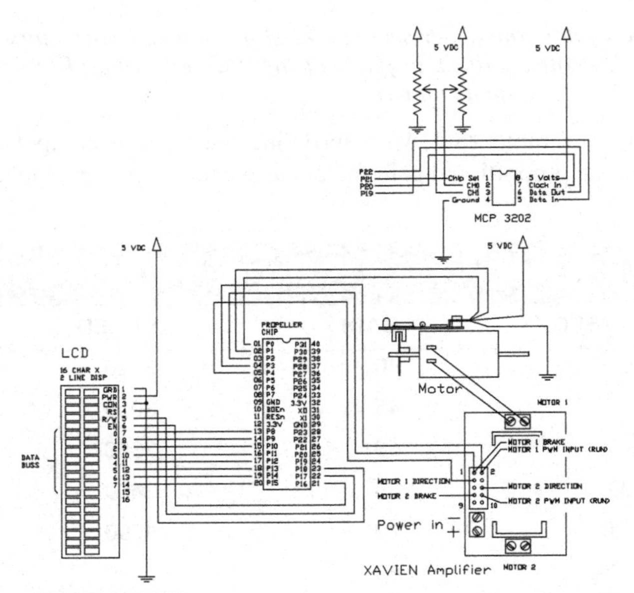

Figure 28-5 Two-potentiometer setup motor-control diagram

Let's add the potentiometers so that we can vary two variables as we play with our programs. The wiring diagram for a setup with two potentiometers is shown in Figure 28-5.

Ramping

First, let's design a simplified but very crude ramp where we learn to run the motor at a couple of speeds under program control. In this situation, the motor follows the following scheme:

1. Stopped.
2. Run at a slow speed (0.5 seconds).
3. Run at a fast speed (1 second).
4. Run at a slow speed (0.5 seconds).
5. Stop for 1 second.
6. Repeat last four steps.

Let's divide the moves into 1/100-second segments so that we will run slow for 50 segments, run fast for 100 segments, run slow for 50 segments, and then stop. This is done by managing the gain over a two-second timeframe. Let's look at just the part of the program that manages the gain (see Program 28-10).

Program 28-10 Program Segment to Figure Gains

```
PUB FigureGain              'power to motor
  repeat                    'loop
    Case time               'base the gain on the time segment number
      0..50:                'first 50 segments
        gain:=20            'slow speed
      51..150:              'next 100 segments
        gain:=100           'fast speed
      151..200:             'next 50 segments
        gain:=20            'slow speed
      201..300:             'next 100 segments
        gain:=0             'Gain=0 means stop for a while
    waitcnt(120_000+cnt)    'define length of each time segment as 0.1 secs
    time:=time+1            'add to time segment number
    if time==300            'check top time value
      time:=0               'reset time counter
```

The complete program listing incorporating this gain algorithm is shown in Program 28-11.

Program 28-11 Run Motor Slow/Fast/Slow/Stop and Then Repeat under Computer Control (No Human Intervention)

```
{{{4 Sep 09   Harprit Sandhu
MotorSloFstSlo.Spini
Propeller Tool Version 1.2.6
Chapter 28  Program 11

This program moves the motor in a slow, fast,slow, stop sequence
to demonstrate the control of the speed over a period of equal time
segments. This is an absolutely minimal ramping demonstration. Only
two speeds are used. This could be expanded to a very complicated
algorithm with lots of gains and speed specifications.

We are not using the encoder or the potentiometers in this program.

Connections are
Amplifier brake       P2
Amplifier PWM         P3
Amplifier direction   P4
Potentiometer        P19
LCD on the usual P12..P18 4 bit mode
Revisions

}}
```

(continued)

Program 28-11 Run Motor Slow/Fast/Slow/Stop and Then Repeat under Computer Control (No Human Intervention) (*continued*)

```
OBJ
  LCD      : "LCDRoutines4" 'for the LCD methods

CON
  _CLKMODE=XTAL1 + PLL2X          'The system clock spec
  _XINFREQ    = 5_000_000

  AmpBrake    =2
  AmpPWM      =3
  MotorDir    =4

VAR
  long     Pos[3]          'Create buffer for encoder
  long stack2[25]          'space for Cog_LCD
  long stack3[25]          'space for Cog_SetMotorPower
  long stack4[25]          'space for Cog_RunMotor
  long stack5[25]          'space for Cog_FigureGain
  long gain
  long time
  long pulswidth

PUB Go
  cognew(cog_LCD, @stack2)
  cognew(SetMotorPower, @stack3)
  cognew(RunMotor(AmpPWM),@Stack4)
  cognew(FigureGain,  @stack5)
  repeat

PRI cog_LCD                             'manage the LCD
  LCD.INITIALIZE_LCD                    'initialize the LCD
  repeat                                'LCD loop
    LCD.POSITION (1,1)                  'Go to 1st line 1st space
    LCD.PRINT(STRING("Time=" )) 'Error
    LCD.PRINT_DEC(time)                  'print the pot reading
    LCD.SPACE(2)                         'erase over old data
    LCD.POSITION (2,1)                  'Go to 2nd line 1st space
    LCD.PRINT(STRING("Gain=" ))          'Potentiometer
    LCD.PRINT_DEC(gain)                  'print the pot reading
    LCD.SPACE(2)                         'erase over old data

PUB SetMotorPower
  dira[AmpBrake..MotorDir]~~            'These pins control the motor amp
  outa[AmpBrake]~                       'turn off the amp brake
  repeat                                'loop
    outa[MotorDir]~                     'move negative
    PULSWIDTH:=gain*450                 'set gain
```

(*continued*)

Program 28-11 Run Motor Slow/Fast/Slow/Stop and Then Repeat under Computer Control (No Human Intervention) (*continued*)

```
PUB RunMotor(Pin)|WaveLength,period     ' toggle the output line, set
  dira[AmpBrake..MotorDir]~~           'g the three amplifier lines
  ctra[30..26]:=%00100                 'Set this cog's "A Counter" to run PWM
  ctra[5..0]:=Pin                      ' "A pin" g to Pin# here
  frqa:=1                              'Set this counter's frqa value to 1
  PulsWidth:=0                         'Start with pulsewidth=0
  WaveLength:=clkfreq/50               ' time for the pulse width to 10 ms
  period:=cnt                          ' current value of the counter
  repeat                               'power PWM routine.
    phsa:=-PulsWidth                   ' high for Pulsewidth counts
    period:=period+WaveLength          'Calculate wave length
    waitcnt(period)                    'Wait for the wavelength

PUB FigureGain                   'power to motor
  repeat                         'loop
    Case time                    'base the gain on the time segment number
      0..100:                    'first 50 segments
        gain:=30                 'slow speed
      101..200:                  'next 100 segments
        gain:=200                'fast speed
      201..300:                  'next 50 segments
        gain:=30                 'slow speed
      301..399:                  'next 100 segments
        gain:=0                  'Gain=0 means stop for a while
    waitcnt(120_000+cnt)         'length of each time segment is 0.1 secs
    time:=time+1                 'add to time segment number
    if time==400                 'check top time value for reset
      waitcnt(clkfreq/4+cnt)  'wait to look at LCD
      time:=0                    'reset time counter
```

A typical, simple, coordinated move for each motor consists of a ramp up in speed, a run at a fixed speed, and then a ramp down in speed. Each move component for each motor takes the same amount of time, so all moves stay in sync. In most cases, it is desirable that the ramps be short, but we will use a longer ramp to start with so that we can actually see what is going on. Let's add a move distance and implement this scheme in the next program.

To keep it in round numbers, let's define the move we will study as follows: exactly 5 seconds to move exactly 5,000 counts of a 42-slot encoder (this was the situation for my particular motor; your situation may be different). Because that is the total specification for the move, we get to figure out what we need to do to get this done. Let's say we will run at the maximum speed for 3 seconds. This will give us 1 second to speed up and 1 second to slow down. Let's agree that we will set up the program to update the gain 100 times a second during the move, for a total of 500 iterations. In our move profile, we will have 100 steps to speed up and 100 to slow down. The other 300 interrupts will be at the constant traverse speed. Because we are ramping up and down at a constant rate, the ramps will each use up 1 second and move a total of 1,000 counts for both ramps (half as much as if they were at full speed for 2 seconds). This means that during the middle 3 seconds, *the motor has to move 4,000 counts*—or 1,333 counts per second. Considerably more than we had in mind! We could compensate for this by shortening the ramping times if that was critical, keeping in mind that a motor will ramp up only so fast and that the load affects this time.

First, let's just ramp up and down. In Program 28-12, Pot1 controls the time between iterations (the larger the setting, the shorter the time and the faster the speed up), and Pot2 controls the distance moved during the ramp up and ramp down. In this program, the motor ramps up, ramps down, delays, and then goes back to the starting position rapidly (to prevent position register overflow). It then repeats the cycle (see Figure 28-6).

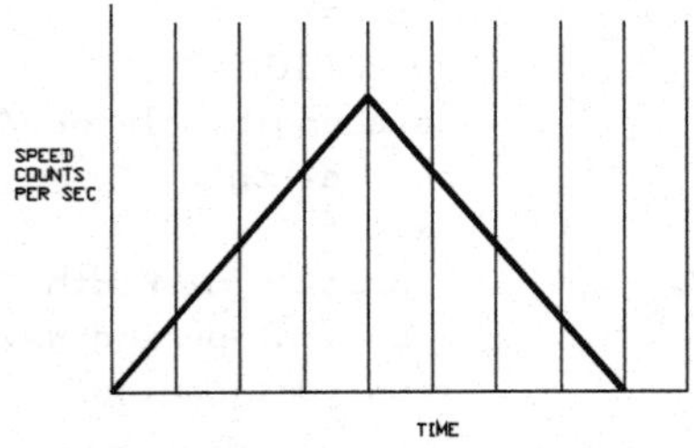

Figure 28-6 Simple ramp up/ramp down

Program 28-12 Ramp Up/Ramp Down Move with Potentiometer-Controlled Variables (Based on Time Segments, Not Encoder Counts)

```
{{04 Nov 09   Harprit Sandhu
MotorBasicMoveAuto.Spin
Propeller Tool Version 1.2.6
Chapter 28  program 12

This program makes a ramped move of a fixed number of encoder counts.
During the move you can change the following:
Pot1 controls the total number of counts in the move.
Pot2 controls the ramp up speed.

We will use the information gained here to design the equations
to set the gain to make the move as fast as possible without
overshoot.

You may have to adjust the parameters for your particular motor.

Connections are
Amplifier brake        P2
Amplifier PWM          P3
Amplifier direction    P4

Potentiometer         P19

LCD on the usual P8..P18

}}
```

(continued)

Program 28-12 Ramp Up/Ramp Down Move with Potentiometer-Controlled Variables (Based on Time Segments, Not Encoder Counts) (*continued*)

```
OBJ
  Encoder : "Quadrature Encoder"
  LCD     : "LCDRoutines4" 'for the LCD methods
  UTIL    : "Utilities"   'for general methods

CON
  _CLKMODE=XTAL1+ PLL2X         'The system clock spec
  _XINFREQ = 5_000_000

VAR
  long     Pos[3]           'Create buffer for encoder
  long stack2[25]           'space for Cog_LCD
  long stack3[25]           'space for Cog_SetMotorPower
  long stack4[25]           'space for Cog_RunMotor
  long stack5[25]           'space for Cog_FigureGain
  long stack6[25]           'space for Cog_Start
  long stack7[25]           'space for readpots
  long pulswidth            '
  long startPosition        '
  long PresentPosition      '
  long TargetPosition       '
  long PositionError        '
  word startFlag            '
  long gain                 '
  long pot1
  long Pot2
  word iters
  word index
  word delay

PUB Go
  cognew(cog_LCD,          @stack2)
  cognew(SetMotorPower,    @stack3)
  cognew(RunMotor(3),      @Stack4)
  cognew(figureGain,       @stack5)
  cognew(StrtFlag,         @stack6)
  cognew(ReadPots,         @stack7)
  Encoder.Start(0, 1, 0, @Pos)       'Reads the encoder position
  startPosition:=Pos[0]              'position read
  repeat                             'movement loop
    iters:=pot2                      'set number of iterations to perform
    index:=0                         'reset index
    repeat iters                     'do iterations
      index:=index+1                 'increment index
      targetPosition:=targetPosition+index    'set new position
      waitcnt(clkfreq/Pot1+cnt)               'wait time for iteration
```

(*continued*)

Program 28-12 Ramp Up/Ramp Down Move with Potentiometer-Controlled Variables (Based on Time Segments, Not Encoder Counts) (*continued*)

```
    repeat iters                        'now do the slow down
      index:=index-1
      targetPosition:=targetPosition+index
      waitcnt(clkfreq/Pot1+cnt)
    repeat while startFlag==0           'wait till done
    waitcnt(clkfreq+cnt)                'delay to see stop
    targetPosition:=startPosition       'set to go back
    waitcnt(24_000+cnt)                 'wait to get done
    repeat while startFlag==0           'wait till done
    waitcnt(clkfreq+cnt)                'delay to see stop

PRI SetMotorPower                       'bases it ~ on positional error
  dira[2..4]~~                          'These pins control the motor amp
  repeat                                'loop
    PresentPosition:=pos[0]             'read the encoder 0 position
    PositionError:=TargetPosition-PresentPosition
    case PositionError                  'decision variable
      -100_000_000..-2:                 'negative range
         outa[4]~~                      'move in position direction
         PULSWIDTH:=gain*450            'set gain
      -1..1 :                           'range for stopping
         PULSWIDTH:=0*450               'gain is 0
      2..100_000_000:                   'positive range
         OUTA[4]~                       'move in negative direction
         PULSWIDTH:=gain*450            'set gain

PRI cog_LCD                             'manage the LCD
  LCD.INITIALIZE_LCD                    'initialize the LCD
  repeat                                'LCD loop
    LCD.POSITION (1,1)                  'Go to 1st line 1st space
    LCD.PRINT(STRING("Tr=" ))           'target position
    LCD.PRINT_DEC(TargetPosition)       'print target
    LCD.SPACE(4)                        'erase over old data
    LCD.POSITION (1,10)                 'Go to 1st line 10 space
    LCD.PRINT_DEC(pot1)                 'print Pot1
    LCD.SPACE(2)                        'erase over old data
    LCD.POSITION (1,14)                 'Go to 1st line 14 space
    LCD.PRINT_DEC(pot2)                 'print Pot2
    LCD.SPACE(2)                        'erase over old data
    LCD.POSITION (2,1)                  'Go to 2nd line 1st space
    LCD.PRINT(STRING("Pp=" ))           'Present position
    LCD.PRINT_DEC(presentposition)      'print position
    LCD.SPACE(4)                        'erase over old data
    LCD.POSITION (2,10)                 'Go to 2nd line 10 space
    LCD.PRINT(STRING("Start=" ))        'start flag
    LCD.PRINT_DEC(StartFlag)            'print flag status
```

(continued)

Program 28-12 Ramp Up/Ramp Down Move with Potentiometer-Controlled Variables (Based on Time Segments, Not Encoder Counts) (*continued*)

```
PRI RunMotor(Pin)|Cycle_time,period  ' toggle the output line, set
  dira[2..4]~~                 'g the three amplifier lines
  ctra[30..26]:=%00100         'Set this cog's "A Counter" to run PWM
  ctra[5..0]:=Pin              'Set the "A pin" of this cog to Pin=3
  frqa:=1                      'Set this counter's frqa value to 1
  PulsWidth:=0                 'Start with position=0
  Cycle_time:=clkfreq/100      'Set the time for the wave L to 10 ms
  period:=cnt                  ' current value of the counter
  repeat                       'power PWM routine.
    phsa:=-PulsWidth           'high pulse for Pulsewidth counts
    period:=period+Cycle_time'Calculate cycle time
    waitcnt(period)            'Wait for end of cycle time

PRI StrtFlag                   'used to determine if motor completed move
  repeat
    if ||PositionError<2       'is move almost done
      startFlag:=1             'if it is it's ok to start
    else                       'if not
      startFlag:=0             'do not start next move

PUB figureGain                 'power to motor
  repeat                       'loop
    case ||PositionError       'pot is NOT READ
      0..20:gain:=10           'in these comparisons
      21..80:gain:=20          'the gain is set by the
      81..200:gain:=60         'amount of the error
      101..80000:gain:=100     'in the position.

PRI Readpots                        'read to 0-4095 and then div by 16
  repeat
    pot1:=UTIL.Read3202_0/16+15    'adding 15 eliminates jerkiness
    pot2:=UTIL.Read3202_1/16
```

Play with Program 28-12 to see how the distance moved and the speed of the move affect the motion we get. Think about how to design an algorithm that ties the gain to the distance to be moved and the speed to get smooth startups and stops. The main thing to notice is that the speed is always a function of the gain, and if you amplify the gain value, the motor gets the move done faster. A relationship exists between the time and the gain multiplier, and how we manage it determines the accuracy of the move timing.

In this program, we are satisfied if the move gets to within two counts of its destination, as determined in the SetMotorPower method. Change that to zero count and see what happens. How does the speed of the move affect this requirement? What can we do to the speed to make this work better at the tail end of each move? Remember, there is no ramp up or ramp down in this program; it's all in how the gain is controlled based on the distance to be moved. (This is a kind of ramp control, but not the managed movement we would need for a complicated coordinated move.)

If you put your finger on the encoder disk to slow it down, you will notice that the increase in the error signal automatically makes the motor power go up to try to keep up with the required speed. However, understand that the motor has now fallen behind its required trajectory. The addition of the integration component into the control algorithm would address this only partially. Before we can have a complete solution, we have to add a time component to the move algorithm so that we have a trajectory based on time. In order to do this, we need to redefine how we are going to make a move.

Now that the gain has been shown to be critical, let's design a program that allows us to play with the gain via a couple potentiometers to see how we might make our moves even more perfect. Let's go back to Program 28-7 and play with it again; then based on what we learn, we will make the appropriate changes to the gain in the program. The program demonstrates that we can ramp up and down and also get a stable motor operation. As long as the stopping is controlled, we can run the rest as fast as we want. However, we will need to control *both* the starting up ramp and the slowing down ramp when we start making coordinated moves.

In Program 28-13, we implement the ramping discussed thus far into a comprehensive move.

Program 28-13 Ramp Up, Run at Constant Rate, Ramp Down, Stop, and Repeat

```
{{4 Sep 09   Harprit Sandhu
MotorSloFstSlo.Spin
Program 13
Propeller Tool Version 1.2.6
Chapter 28 Program 13

This program moves the motor in a slow, fast ,slow, stop sequence
to demonstrate the control of the speed over a period of equal time
segments. This is an absolutely minimal ramping demonstration. Only
two speeds are used. This could be expanded to a very complicated
algorithm with lots of gains and speed specifications.

We are not using the encoder or the potentiometers in this program.

Connections are
Amplifier brake       P2
Amplifier PWM         P3
Amplifier direction   P4

Potentiometer        P19

LCD on the usual P12..P18 4 bit mode

Revisions

}}
```

(continued)

Program 28-13 Ramp Up, Run at Constant Rate, Ramp Down, Stop, and Repeat (*continued*)

```
OBJ
  LCD      : "LCDRoutines4" 'for the LCD methods

CON
  _CLKMODE=XTAL1 + PLL2X          'The system clock spec
  _XINFREQ    = 5_000_000

  AmpBrake    =2
  AmpPWM      =3
  MotorDir    =4

VAR
  long     Pos[3]           'Create buffer for encoder
  long stack2[25]           'space for Cog_LCD
  long stack3[25]           'space for Cog_SetMotorPower
  long stack4[25]           'space for Cog_RunMotor
  long pulswidth            '
  long TimeSegs             '
  long RmpUp                '
  long RmpDn                '
  long RunCn                '
  long move                 '
  long initGain             '
  byte gainStep             '
  long gain                 '

PUB Go                                '
  cognew(cog_LCD, @stack2)            '
  cognew(SetMotorPower, @stack3)      '
  cognew(RunMotor(AmpPWM),@Stack4)    '
  Move:=400                           '
  TimeSegs:=0                         '
  RmpUp:=move/3                       '
  RmpDn:=RmpUp                        '
  RunCn:=MOve-RmpDn-RmpUp             '
                                      '
  Gainstep:=3                         '
  initGain:=15                        '
  gain:=initGain                      '
  repeat                        'loop
    Case timeSegs               'base the gain on the time segment number
      0..RmpUp:                 'first segments
        gain:=gain+gainStep     'slow speed

      RmpUp+1..RmpUp+RunCn:     'next segments
                                'constant speed
```

(continued)

Program 28-13 Ramp Up, Run at Constant Rate, Ramp Down, Stop, and Repeat (*continued*)

```
      RmpUp+RunCn+1..RmpUp+RunCn+RmpDn:    'next segments
        gain:=gain-gainStep    'slow speed

      RmpUp+RunCn+RmpDn+1..RmpUp+RunCn+RmpDn+100:  'next segments
        gain:=0                 'Gain=0 means stop for a while

    timeSegs:=timeSegs+1
    waitcnt(120_000+cnt)   'define length of each time segment as 0.1 secs
    if timeSegs==RmpUp+RunCn+RmpDn+50   'check top time value for reset
      waitcnt(clkfreq+cnt) 'wait to look at LCD
      timeSegs:=0               'reset time counter
      gain:=initGain

PRI cog_LCD                              'manage the LCD
  LCD.INITIALIZE_LCD                     'initialize the LCD
  repeat                                 'LCD loop
    LCD.POSITION (1,1)                   'Go to 1st line 1st space
    LCD.PRINT(STRING("Time=" ))          'Error
    LCD.PRINT_DEC(timeSegs)              'print the pot reading
    LCD.SPACE(2)                         'erase over old data
    LCD.POSITION (2,1)                   'Go to 2nd line 1st space
    LCD.PRINT(STRING("Gain=" ))          'Potentiometer
    LCD.PRINT_DEC(gain)                  'print the pot reading
    LCD.SPACE(4)                         'erase over old data

PUB SetMotorPower
  dira[AmpBrake..MotorDir]~~             'These pins control the motor amp
  outa[AmpBrake]~                        'turn off the amp brake
  repeat                                 'loop
    outa[MotorDir]~                      'move negative
    PULSWIDTH:=gain*450                  'set gain

PUB RunMotor(Pin)|Cycle_time,period    ' toggle the output line, set
  dira[AmpBrake..MotorDir]~~   'g the three amplifier lines
  ctra[30..26]:=%00100         'Set this cog's "A Counter" to run PWM
  ctra[5..0]:=Pin              ' "A pin" of this cog to Pin# here
  frqa:=1                      'Set this counter's frqa value to 1
  PulsWidth:=0                 'Start with pulsewidth=0
  Cycle_time:=clkfreq/50       ' time for the pulse width to 10 ms
  period:=cnt                  'Store the current value of the counter
  repeat                       'power PWM routine.
    phsa:=-PulsWidth           ' high pulse for Pulsewidth counts
    period:=period+Cycle_time  'Calculate cycle time
    waitcnt(period)            'Wait for the cycle time
```

Program 28-13 gives us everything we need to ramp up, run at a constant speed, and then ramp down. However, the scheme is defined by the number of time segments

(TimeSegs), each of which represents one iteration of the moves in the Go method. The program move can be totally defined by the variables set at the top of the Go method. This program needs the following refinements to make it more useful:

1. Convert to a time-based regime.
2. Set a limit on the top gain to be allowed.
3. Add the ability to change the ramping times relative to the total time.
4. Convert to the number of encoder counts moved within the time specification.

The flexibility that can be added is limited only by your imagination. For experimental and investigative purposes, one thing we could do immediately is to make certain variables dependent on the values of the two potentiometers we have available. The ramping we have been discussing is illustrated in Figure 28-7.

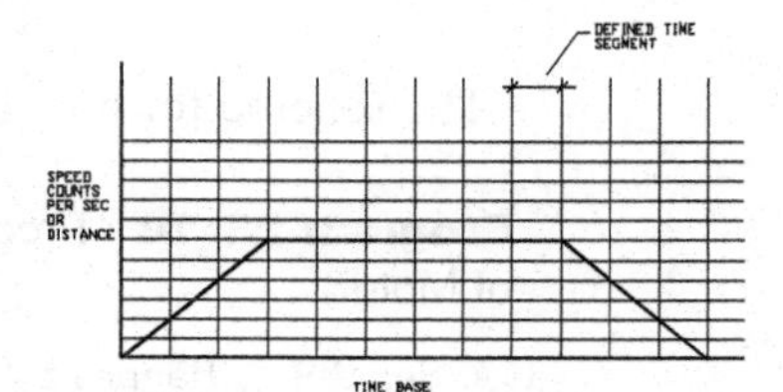

Figure 28-7 **The speed/time path to be followed by the motor**

Look at the PWM signal on the oscilloscope as you run this program. You can see it change as the motor goes through the cycle. In this particular experiment, it gets to above a 50% duty cycle, which is not desirable because this does not leave enough room to provide for sudden load changes. As discussed earlier, under no load, the motor should be able to follow the profile with no more than a 35% to 50% PWM duty cycle.

R/C Signal Use

A lot of the work we do with motors as hobbyists (and this is useful for other applications too) has to do with running them from the signal we get from a hobby R/C receiver/transmitter. This is no different from running the motor from a PWM signal. The math we do has to convert the R/C signal pulses into what we need for running a motor. Let's take a closer look at this with the radio control signal in mind. Assume that these signals come in at pulses between 750 and 2,250 ms with zero motion at 1,500 ms. We want this signal to be converted to a number between –127 and +127 and fed to the instruction to run a motor.

The pseudocode for doing this is as follows:

```
Read the signal pulse length from the radio receiver (500 to 2,500 ms).
Convert it to a number going from -127 to 0 to +127.
Multiply by two to get full-scale operation.
Set the direction of the motor from the value sign.
Run or position the motor.
Loop, and do it again.
```

We can also feed the value read to the target position and use it that way, just as we have been doing in the earlier programs. The caveat would be that the target value in the motor position counter underflows at 0 and overflows at 16.7 meg, and you have to reset the encoder position counter before this happens.

To turn a DC servo motor to a position-seeking servo based on an R/C signal, we have to do the following things:

1. Read the pulse width from the radio.
2. Divide to get a stable number to provide a half revolution like an R/C servo.
3. Position the motor to the radio signal.
4. Repeat.

The scheme for doing this is implemented in Program 28-14.

Program 28-14 Program to Turn Motor into an R/C Servo (Remote Position Control of Motor)

```
{14 Sep 09   Harprit Sandhu
MotorPositionRC.Spin
Propeller Tool Version 1.2.6
Chapter 28  Program 14

This program Positions the DC motor from a hobby R/C transmitter
just like it would a R/C hobby servo, but many revs.

You may have to adjust the parameters for your particular motor.

Connections are
Amplifier brake       P2
Amplifier PWM         P3
Amplifier direction   P4
Radio signal in       P24
LCD on the usual      P8..P18
}}

OBJ
  Encoder : "Quadrature Encoder"
  LCD     : "LCDRoutines4" 'for the LCD methods
  UTIL    : "Utilities"  'for general methods

CON
  _CLKMODE=XTAL1+ PLL2X   'The system clock spec
  _XINFREQ = 5_000_000
  BRK      =2
  PWM      =BRK+1
  DIR      =BRK+2
  PulseIn  =24
```

(continued)

Program 28-14 Program to Turn Motor into an R/C Servo (Remote Position Control of Motor) (*continued*)

```
VAR
  long      Pos[3]           'Create buffer for encoder
  long stack2[35]            'space for Cog_LCD
  long stack3[35]            'space for Cog_SetMotorPower
  long stack4[35]            'space for Cog_RunMotor
  long stack5[35]            'space for Cog_Encoder
  long startcycle            '
  long endPulse              '
  long endCycle              '
  long CycleLen              '
  long PulseLen              '
  long pulswidth
  long freq                  '
  long MotPwr                '
  long PresPos
  long TargPos

PUB Go
  cognew(cog_LCD,        @stack2)
  cognew(SetMotorPower, @stack3)
  cognew(PosMotor(3),    @Stack4)
  cognew(Enco,           @Stack5)
  Encoder.Start(0, 1, 0, @Pos)        'Read the encoder position
  Dira[PulseIn]~
  repeat
    waitpeq(|<PulseIn, |<PulseIn, 0)  'wait for line to go low.
    waitpeq(0, |<PulseIn, 0)          'wait for line to go low.
    waitpeq(|<PulseIn, |<PulseIn, 0)  'wait for line to go low.
    startCycle:=CNT                   'read the timer count 0)
    waitpeq(0, |<PulseIn, 0)          'wait for line to go low...
    endPulse:=CNT              'read the timer count for second time
    waitpeq(|<PulseIn, |<PulseIn, 0)  'wait for line to go low.
    endCycle:=CNT              'read the timer count for second time
    pulswidth:=(endPulse-startCycle)  'figure the pulse
    freq:=clkfreq/(endCycle- startCycle)  'figure the freq


PRI cog_LCD                          'manage the LCD
  LCD.INITIALIZE_LCD                 'initialize the LCD
  repeat                             'LCD loop
    LCD.POSITION (1,1)               'Go to 1st line 1st space
    LCD.PRINT(STRING("POSN=" ))      'POSITION
    LCD.PRINT_DEC(Prespos)           'print position
    LCD.SPACE(2)                     'erase over old data
```

(*continued*)

Program 28-14 Program to Turn Motor into an R/C Servo (Remote Position Control of Motor) (*continued*)

```
    LCD.POSITION (1,11)             'Go to 1st line 11 space
    LCD.PRINT(STRING("DIR=" ))      'Direction of motor power
    LCD.PRINT_DEC(ina[Dir])         'print
    LCD.SPACE(2)                    'erase over old data
    LCD.POSITION (2,1)              'Go to 2nd line 1st space
    LCD.PRINT(STRING("TARG=" ))     'TARGET
    LCD.PRINT_DEC(TargPos)          'print
    LCD.SPACE(2)                    'erase over old data }

PUB SetMotorPower
  dira[BRK..DIR]~~                  'These pins control the motor amp
  repeat                            '
    targPos:=pulswidth/10           'loop
    if TargPos<PresPos              '
       OUTA[DIR]~~                  '
       dira[BRK]~                   '
       MotPwr:=||(TargPos-PresPos)*200
    if TargPos==PresPos             '
       dira[BRK]~~                  '
       MotPwr:=0                    '
    if TargPos>PresPos              '
       OUTA[DIR]~                   '
       dira[BRK]~                   '
       MotPwr:=4|(TargPos-PresPos)*200

PUB PosMotor(Pin)|WaveLength,period   ' toggle the output line
  dira[BRK..DIR]~~                  ' the three amplifier lines
  ctra[30..26]:=%00100              'Set this cog's "A Counter" to run PWM
  ctra[5..0]:=Pin                   ' "A pin" of this cog to Pin# here
  frqa:=1                           'Set this counter's frqa value to 1
  motpwr:=0                         'Start with pulsewidth=0
  WaveLength:=clkfreq/100           ' time for the pulse width  10 ms
  period:=cnt                       'Store the current value of the counter
  repeat                            'power PWM routine.
    phsa:=-MotPwr                   ' high pulse for Pulsewidth counts
    period:=period+WaveLength       'Calculate wave length
    waitcnt(period)                 'Wait for the wavelength

PRI Enco                            'reads encoder
  dira[0..1]~                       'encoder pins are inputs
  repeat
    PresPos:=Pos[0]
```

In Program 28-14, we are using the waitpeq command to wait for the signals coming in on the line we are monitoring. This is the preferred way to wait for a line to change. Another way of doing this is shown in Program 28-15. Compare the two methods.

Finally, if you want to control the speed of the motor from an R/C signal, the pulse width read from the radio has to be converted to a number between 0 and 255. This number is then used just like we used the signal from the potentiometer to run the motor. The code we are using was originally discussed in Chapter 24 on hobby servos.

The pseudocode for running the motor from an R/C signal is as follows:

```
Read the incoming pulse width in microseconds.
Convert the range from 750 to 2,250 microseconds to a value from -127 to +128.
Use the converted value to determine the motor direction.
Use the converted value to determine the motor gain.
Run the motor.
Read the incoming pulse width again.
```

The conversion can be made as follows:

Motor gain = (pulsewidth in – 1,500) * 255/1,500

We can assign cogs to this problem's solution as follows:

- **Cog_0** Start the other cogs, set up I/O, and loop to read input pulse length.
- **Cog_1** Manage the LCD display.
- **Cog_2** Calculate motor gain as a pulse width.
- **Cog_3** Run the motor.

The code for these cogs is shown in Program 28-15.

Program 28-15 Program in Which Motor Speed and Direction Follow an R/C Servo Signal (Remote Speed Control of a Motor)

```
{{14 Sep 09   Harprit Sandhu
MotorSpeedDirRC.Spin
Propeller Tool Version 1.2.6
Chapter 28  Program 15

This program runs the motor back and forth from a
signal received from a hobby R/C transmitter as would
be used by a servo.

You may have to adjust the parameters for your particular motor.

Connections are
Amplifier brake        P2
Amplifier PWM          P3
Amplifier direction    P4
Radio signal in        P24
LCD on the usual       P8..P18
}}
```

(continued)

Program 28-15 Program in Which Motor Speed and Direction Follow an R/C Servo Signal (Remote Speed Control of a Motor) (*continued*)

```
OBJ
  Encoder : "Quadrature Encoder"
  LCD     : "LCDRoutines4" 'for the LCD methods
  UTIL    : "Utilities"  'for general methods

CON
  _CLKMODE=XTAL1+ PLL2X    'The system clock spec
  _XINFREQ = 5_000_000
  BRK       =2
  PWM       =BRK+1
  DIR       =BRK+2
  RadioIn   =24

VAR
  long      Pos[3]          'Create buffer for encoder
  long stack2[25]           'space for Cog_LCD
  long stack3[25]           'space for Cog_SetMotorPower
  long stack4[25]           'space for Cog_RunMotor
  long startcycle           '
  long endPulse             '
  long endCycle             '
  long CycleLen             '
  long PulseLen             '
  long freq                 '
  long MotPwr               '

PUB Go
  cognew(cog_LCD,        @stack2)
  cognew(SetMotorPower, @stack3)
  cognew(RunMotor(3),@Stack4)
  Dira[RadioIn]~
  repeat
    repeat while ina[RadioIn]==1  'wait for line 1 to go low.
                                  'make sure that we see
                                  'at full wave when we start measuring
    repeat while ina[RadioIn]==0  'wait for line 1 to go low. Manual
    startCycle:=CNT               'read the timer count

    repeat while ina[RadioIn]==1  'wait for line 1 to go hi. See Manual
    endPulse:=CNT                 'read the timer count for second time

    repeat while ina[RadioIn]==0  'wait for line 1 to go low. Manual.
    endCycle:=cnt

    PulseLen:=(endPulse-startCycle)/10  'figure the pulse
    CycleLen:=endCycle-startCycle       'figure the cycle Len
    freq:=10_000_000/CycleLen           'figure the freq
```

(continued)

Program 28-15 Program in Which Motor Speed and Direction Follow an R/C Servo Signal (Remote Speed Control of a Motor) *(continued)*

```
PRI cog_LCD                             'manage the LCD
  LCD.INITIALIZE_LCD                    'initialize the LCD
  repeat                                'LCD loop
    LCD.POSITION (1,1)                  'Go to 1st line 1st space
    LCD.PRINT(STRING("PWR=" ))          'Power
    LCD.PRINT_DEC(MotPwr)               'print power
    LCD.SPACE(2)                        'erase over old data
    LCD.POSITION (1,11)                 'Go to 1st line 11 space
    LCD.PRINT(STRING("DIR=" ))          'Direction
    LCD.PRINT_DEC(ina[Dir])             'print
    LCD.SPACE(2)                        'erase over old data
    LCD.POSITION (2,1)                  'Go to 2nd line 1st space
    LCD.PRINT(STRING("WLN=" ))          'Cycle length
    LCD.PRINT_DEC(CycleLen)             'print
    LCD.SPACE(2)                        'erase over old data
    LCD.POSITION (2,11)                 'Go to 2nd line 11 space
    LCD.PRINT(STRING("FRQ=" ))          'Freq
    LCD.PRINT_DEC(freq)                 'print
    LCD.SPACE(2)                        'erase over old data

PUB SetMotorPower
  dira[BRK..DIR]~~                        'These pins control the motor amp
  repeat                                  'loop
     case pulselen                        '
       300..1500:                         'low values
         outa[DIR]~                       'Motor dir low
         MotPwr:=1500-pulselen            'reset value
       1501..3000:                        'high values
         outa[DIR]~~                      'Motor dir hi
         MotPwr:=pulselen-1500            'reset value

PUB RunMotor(Pin)|WaveLength,period    ' toggle the output line, set
  dira[2..4]~~                            ' the three amplifier lines
  ctra[30..26]:=%00100                 'Set this cog's "A Counter" to run PWM
  ctra[5..0]:=Pin                      'Set the "A pin" of this cog to Pin
  frqa:=1                              'Set this counter's frqa value to 1
  PulseLen:=0                          'Start with position=0
  WaveLength:=clkfreq/100              ' time for the pulse width to 10 ms
  period:=cnt                          'Store the current value of the counter
  repeat                               'power PWM routine.
    phsa:=(-MotPwr+70)*50              'S high pulse for Pulsewidth counts
    period:=period+WaveLength          'Calculate wave length
    waitcnt(period)                    'to end of wave
```

In Program 28-15, we are waiting for the signals to change as read by the "repeat while" loops. Compare this scheme with how we read the signals in Program 28-14.

Other schemes can be constructed as may be necessary. Just modifying the techniques that have been demonstrated here. They will work for most applications.

Do not forget that if you want to create a scheme in which the target is fed by the error signal, you have to accommodate any condition that may underflow or overflow the 31 bit position register from time to time.

Some Advanced Considerations You Should Be Aware Of

When you want to run a complicated profile-following program, the technique used is complicated. First, the moves are described in the EIA RS-274 language standard. This is the language used by CNC machines and consists of G codes, M codes, and so on, followed by X, Y, and Z positioning commands. This program is interpreted as a data source that is used to create a series of pulses for as many axes as are needed. These pulses are then fed to each axis in the operating system. I understand that it is necessary to read the database five steps ahead of the step being executed to make sure that the current move will not be affected by a future move owing to tool compensations and move melding, etc. In what I have described in this chapter, we are a long way from being able to do this, and it is beyond the capability of the Propeller in Spin. In other words, you have to create a program that reads the first five lines in the RS-274D data, converts the instructions into pulses, feeds the first set of pulses to the appropriate axes, and then reads the next instruction and executes it, always reading five steps ahead of the execution. It's not simple. However, simpler schemes that serve a one- or two-axis system may well do what we need done.

The other hard part is the melding of one instruction into the next. You cannot stop at the end of each instruction in many applications. The machine must keep moving seamlessly. The schemes for doing this are quite complicated and beyond what we need to understand at the beginner's level, but we should be aware of the problem. Something to think about!

29

RUNNING SMALL AC MOTORS: CONTROLLING INDUCTIVE LOADS

Oftentimes it is necessary for the engineer to control a small fractional horsepower electrical motor as a part of his experimental apparatus. The easiest way to do this is to use a solid-state relay that will accept a TTL signal as its input and have the capacity to control the current and voltage needed by the motor.

A number of vendors provide very easy-to-use, solid-state relays that meet these specifications, via the Internet. Figure 29-1 shows the relay manufactured by Crydom and available from Jameco.

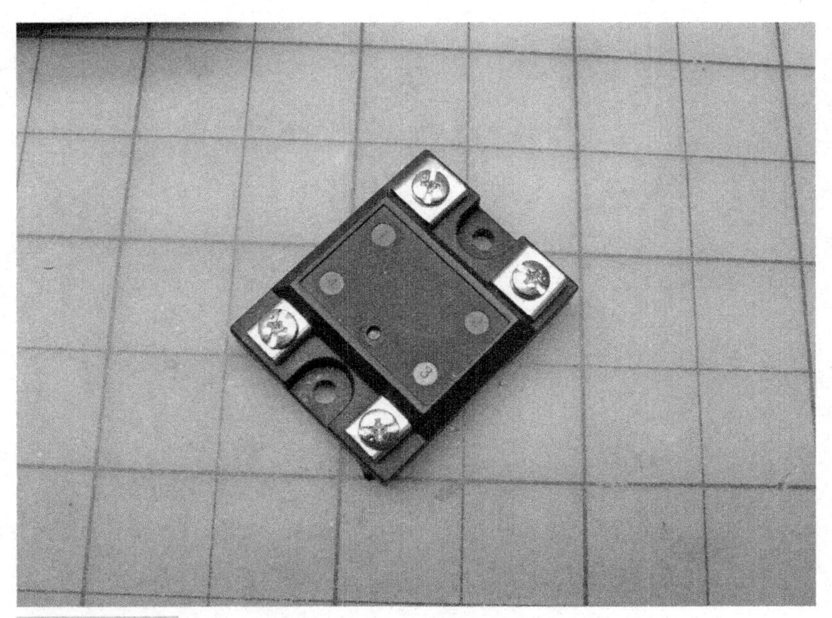

Figure 29-1 A solid-state relay (note the indicator LED)

Check to see that the relay will handle an AC load at least twice the current you want to control. Electric motors should not be turned on and off more than a couple of times a minute and can usually *not* use a PWM signal.

The control of inductive loads is similar to the control of electric motors. Provide a diode across the inductance to absorb the back EMF when the electric field in the inductance collapses. Size the diode so that it can amply handle the back current and voltage created by the inductance shutoff. Over-sizing the diode by a factor of three or four is recommended.

If you need to control a larger motor, you can do so by controlling the starter for the relay with a solid-state relay. Using a relay as an intermediate device expands the possibilities tremendously, but does not allow the modulation of signals. In the usual state, larger loads need to be started no more than a few times in a minute. See the guidelines provided in the National Electrical Code for more information.

If you have to control a reversing load, it may be necessary to use two solid-state relays—one for each direction—and to provide software interlocks and delays to ensure proper and safe operations.

Safety becomes a major consideration when controlling large loads. Get the help of an expert when managing loads you are not familiar with.

The requirements of the National Electrical Code have to be met when using more than 24 volts. An electrician can help you meet the requirements. Local codes will also need to be followed, and your company may have requirements that have to be met as well.

Part IV

APPENDIXES

LCDRoutines4 AND UTILITIES OBJECT LISTINGS

The listings for the LCDRoutines4 and Utilities objects are provided in this appendix in case this information ceases to be available on the Internet for any reason in the future. The LCDRoutines4 and Utilities objects are called from almost all the methods in this book, so they are essential for running almost all the programs in this book.

These listings can be downloaded from the Encodergeek.com website for the foreseeable future. You can cut and paste the files from there. Because indentation is an important part of a program listing in Spin, the source you use must preserve the indentation.

```
{{21 Sep 09     Harprit Sandhu
LCDRoutines4.spin
Propeller Tool Version 1.2.6
Appendix Program 01

LCD ROUTINES for a 4 bit data path.

The following are the names of the methods described in this program

  INITIALIZE_LCD
  PRINT (the_line)
  POSITION (LINE_NUMBER, HOR_POSITION) | CHAR_LOCATION
  SEND_CHAR (DISPLAY_CHAR)
  SEND_CHAR (DISPLAY_CHAR)
  PRINT_DEC (VALUE) | TEST_VALUE
  PRINT_HEX (VALUE, DIGITS)
  PRINT_BIN (VALUE, DIGITS)
  CLEAR
  HOME
  SPACE (QTY)
```

```
Revisions
  04 Oct 09  Initialize made more robust, misc unnecessary calls removed.

}}
CON                                'all the constants used by all the METHODS
                                   'in this program have to be listed here
  _CLKMODE=XTAL1 + PLL2X           'The system clock spec. 2 X multiplier
  _XINFREQ   = 5_000_000           'ext crystal is 5 MHz, so 10 MHz operation
  DataBit4   = 12                  'are named so that they can be called by
  DataBit5   = 13                  'name if the need ever arises
  DataBit6   = 14                  '
  DataBit7   = 15                  '
  RegSelect  = 16                  'The three control lines
  ReadWrite  = 17                  'The three control lines
  Enable     = 18                  'The three control lines
  high       =1                    'define the High state
  low        =0                    'define the Low state
  Inv_high   =0                    'define the Inverted High state
  Inv_low    =1                    'define the Inverted Low state

VAR                                'these are the variables we will use
   byte  temp                      'for use as a pointer
   byte  index                     'to count characters

PUB Go
  INITIALIZE_LCD
  repeat
    print(String("4bit mode line 1"))
    position(2,1)
    print(String("4bit mode line 2"))
    waitcnt(clkfreq/4+cnt)
    clear
    waitcnt(clkfreq/4+cnt)

{{initialize the LCD to use 4 lines of data
Includes the half second delay, clears the display and positions to 1,1
no variables used
}}
PUB INITIALIZE_LCD             'The addresses and data used here are
  waitcnt(150_000+cnt)         'specified in the Hitachi data sheet for the
  DIRA[DataBit4..Enable]~~     'display. YOU MUST CHECK THIS FOR YOURSELF.
  SEND_INSTRUCTION (%0011)     'Send 1st
  waitcnt(49_200+cnt)          'wait
  SEND_INSTRUCTION (%0011)     'Send 2nd
  waitcnt(1_200+cnt)           'wait
  SEND_INSTRUCTION (%0011)     'Send 3rd
  waitcnt(12_000+cnt)          'wait
  SEND_INSTRUCTION (%0010)     'set for 4 bit mode
  waitcnt(12_000+cnt)          'wait
```

```
  SEND_INSTRUCTION2 (%0010_1000)  'Sets DL=4 bits, N=2 lines, F=5x7 font
 ' waitcnt(12_000+cnt)            'wait
  SEND_INSTRUCTION2 (%0000_1110)  'Display on, Blink off, Sq Cursor off
 ' waitcnt(12_000+cnt)            'wait
  SEND_INSTRUCTION2 (%0000_0110)  'Move Cursor, Do not shift display
 ' waitcnt(12_000+cnt)            'wait
  SEND_INSTRUCTION2 (%0000_0001)  'clears the LCD
  'waitcnt(12_000+cnt)            'wait
  POSITION (1,1)

{{Sends instructions as opposed to a character to the LCD
no variables are used
}}
PUB SEND_INSTRUCTION (D_DATA)   'set up for writing instructions
  CHECK_BUSY                    'wait for busy bit to clear before sending
  OUTA[ReadWrite] := 0          'Set up to read busy bit
  OUTA[RegSelect] := 0          'Set up to read busy bit
  OUTA[Enable]    := 1          'Set up to toggle bit H>L
  OUTA[DataBit7..DataBit4] := D_DATA 'Ready to READ data in
  OUTA[Enable]    := 0                'Toggle the bit H>L to Xfer the data

{{Sends an instruction as opposed to a character to the LCD
no variables are used
}}
PUB SEND_INSTRUCTION2 (D_DATA)   'set up for writing instructions
  CHECK_BUSY                     'wait for busy bit to clr before sending
  OUTA[ReadWrite] := 0           'Set up to read busy bit
  OUTA[RegSelect] := 0           'Set up to read busy bit
  OUTA[Enable]    := 1           'Set up to toggle bit H>L
  OUTA[DataBit7..DataBit4] := D_DATA>>4   'Ready to READ data in
  OUTA[Enable]    := 0               'Toggle the bit H>L to Xfer the data
  OUTA[Enable]    := 1               'Set up to toggle bit H>L
  OUTA[DataBit7..DataBit4] := D_DATA   'Ready to READ data in
  OUTA[Enable]    := 0                'Toggle the bit H>L to Xfer the data

{{Sends a single character to the LCD in two halves
}}
PUB SEND_CHAR (D_CHAR)    'set up for writing to the display
  CHECK_BUSY              'wait for busy bit to clear before sending
  OUTA[ReadWrite] := 0    'Set up to send data
  OUTA[RegSelect] := 1    'Set up to send data
  OUTA[Enable]    := 1    'go high
  OUTA[DataBit7..DataBit4] := D_CHAR>>4  'Send high 4 bits
  OUTA[Enable]    := 0                   'Toggle the bit H>L
  OUTA[Enable]    := 1                   'go high again
  OUTA[DataBit7..DataBit4] :=D_CHAR      'send low 4 bits
  OUTA[Enable]    := 0                   'Toggle the bit H>L
```

```
{{Print a line of characters to the LCD
uses variables index and temp
}}
PUB PRINT (the_line)   'This routine handles more than one Char at a time
                       'called as PRINT(string("the_line"))
                       '"the_line" contains the pointer to line. Line is
                       'because we have to point to the line
                       'zero terminated but we will not use that.  We will
                       'use the string size instead. Easier to understand
  index:=0             'Reset the counter we are using to count chars sent
  repeat               'repeat for all chars in the list
    temp:= byte[the_line][index++] ' char/byte pointed by index
    SEND_CHAR (temp)             'send the 'pointed toÐ char to the LCD
  while index<strsize(the_line)    'keep doing till the last char sent

{{Position cursor
}}
PUB POSITION (LINE_NUMBER, HOR_POSITION) | CHAR_LOCATION  'Pos the cursor  'Line
Number : 1 to 4
  'Horizontal Position : 1 to 20          'specified by the two numbers
  CHAR_LOCATION := (LINE_NUMBER-1) * 64   'figure location. See Hitachi
  CHAR_LOCATION += (HOR_POSITION-1) + 128 'figure location. See Hitachi
 ' CHAR_LOCATION += (HOR_POSITION-1) + 128
 ' CHAR_LOCATION += (HOR_POSITION-1) + 128
  SEND_INSTRUCTION2 (CHAR_LOCATION)       'send instr to position cursor

{{Check for busy
}}
PUB CHECK_BUSY | BUSY_BIT                'routine to check busy bit
  OUTA[ReadWrite] := 1                   'Set to read the busy bit
  OUTA[RegSelect] := 0                   'Set to read the busy bit
  DIRA[DataBit7..DataBit4] := %0000      'Set the entire port to be an input
  REPEAT                          'Keep doing it till clear
    OUTA[Enable]  := 1  'set to 1 to get ready to toggle H>L this bit
    BUSY_BIT := INA[DataBit7]     'the busybit is bit 7 of the byte read
                          'INA is the 32 input pins on the PROP and we
                          'are reading data bit 7 which is on pin 15!
    OUTA[Enable]  := 0    'make the enable bit go low for H>L toggle
  WHILE (BUSY_BIT == 1)   'do it as long as the busy bit is 1
  DIRA[DataBit7..DataBit4] := %1111 'done, so set port back to outputs

{{Print a decimal value, whole numbers only
}}
PUB PRINT_DEC (VALUE) | TEST_VALUE   'for print vals in decimal format
  IF (VALUE < 0)                  'if it is a negative value
    -VALUE                        'change it to a positive
    SEND_CHAR("-")                'and print a - sign on the LCD
```

```
  TEST_VALUE := 1_000_000_000     'individual digits by comparing to this
                              'value  then divide by 10 to get next value
  REPEAT 10                     'There are 10 digits maximum in our system
    IF (VALUE => TEST_VALUE) 'see if our number is bigger than testValue
      SEND_CHAR(VALUE / TEST_VALUE + "0") 'if it is, divide to get the digit
      VALUE //= TEST_VALUE  'figure the next value for the next digit
      RESULT~~              'result of what just did pass it on below
    ELSEIF (RESULT OR TEST_VALUE == 1) 'if t a 1 then div was even
      SEND_CHAR("0")                 'so we sent out a zero
    TEST_VALUE /= 10                 'divide by 10 to test the next digit

{{Print a Hexadecimal value
}}
PUB PRINT_HEX (VALUE, DIGITS)       'for printing values in HEX format
  VALUE <<= (8 - DIGITS) << 2       'you can specify up to 8 digits
  REPEAT DIGITS                     'do each digit
    SEND_CHAR(LOOKUPZ((VALUE <-= 4) & $F : "0".."9", "A".."F"))
                                    'use lookup table to select character

{{Print a Binary value
}}
PUB PRINT_BIN (VALUE, DIGITS)      'for printing values in BINARY format
  VALUE <<= 32 - DIGITS            '32 binary digits is the max for our sys
  REPEAT DIGITS                    'Repeat for each digit desired
    SEND_CHAR((VALUE <-= 1) & 1 + "0") 'send a 1 or a 0

{{Clear screen
}}
PUB CLEAR                           'Clear the LCD display and go home
  SEND_INSTRUCTION2 (%0000_0001)   'This is the clear screen and go home command

{{Go to position 1,1   Does not clear the screen
}}
PUB HOME                            'go to position 1,1.
  SEND_INSTRUCTION2 (%0000_0011)    'Not cleared

{{Print spaces
}}
PUB SPACE (qty)                     'Prints spaces, for between numbers
  repeat (qty)
    PRINT(STRING(" "))



{{21 Sep 09    Harprit Sandhu
Utilities.spin
Propeller Tool Ver 1.2.6
Appendix Program 02
```

```
Program UTILITIES
  Flash          flashes a pin once, toggles it slowly
  Pause          pause in milliseconds
  GetPotValue    returns PotValue  0-255
  GetPotValue2   returns PotValue  0-255
  Read3202_0     reads Channel 0 on the 3202, 12 bits
  Read3202_1     reads Channel 1 on the 3202, 12 bits

  Since the pot readers are an important part of
  the utilities, they are actually run if you run this program.
  Using two methods allows you to use two different pot values,
  one for each pot.

Revisions
21 Sep  09   Set min and max for Pot readings, reduced delay div
             from 2920 to 2900
             Displays both pots and both delays now.
             Using two routines to read the pots allows two pots
             with different values to be set used. Have to be set
             up. Currently they are identical.
             New Xtal changes made.

14 Nov 09    Added potvalue2 to read second pot.
             Added ReadChannel,X to read any of 2 channels from
             a MCP 3202

14 Nov 09    Added independent routines to read the two lines of
             A2D on the MCP3202.

Error Reporting:
Please report errors to harprit.sandhu@gmail.com

}}
CON
  _CLKMODE=XTAL1 + PLL2X         'The system clock spec
  _XINFREQ = 5_000_000           'the oscillator frequency

  high          =1
  low           =0
  PotLine       =19
  PotLine2      =20
  repval        =2
  repval2       =2
  BitsRead      =12
  BitsRead2     =12
  chipSel  = 19
  chipClk  = chipSel+1
  chipDout = chipSel+2
  chipDin  = chipSel+3
```

```
VAR
  long  startCnt
  long  startCnt2
  long  endCount
  long  endCount2
  long  delay
  long  delay2
  long  PotValue
  long  PotValue2
  long  ValuTotal
  long  ValuTotal2
  word  PotReading1
  word  PotReading2
  word  DataRed

OBJ                                 'The methods we will need
  LCD  : "LCDRoutines4"             'for controlling the LCD

PUB Go
  LCD.INITIALIZE_LCD               'set up the LCD
  repeat
    LCD.POSITION (1,1)             'Go to 1st line 1st space
    LCD.PRINT(STRING("Pot1="))     'Potentiometer position
    LCD.PRINT_DEC(PotReading1)     'print pot value
    LCD.SPACE(3)                   'erase over old data

    LCD.POSITION (2,1)             'Go to 1st line 1st space
    LCD.PRINT(STRING("Pot2="))     'Potentiometer position
    LCD.PRINT_DEC(PotReading2)     'print pot value
    LCD.SPACE(3)                   'erase over old data

    Read3202_0                     'Read Pot routine
    Read3202_1                     'Read Pot routine2

PUB FLASH (color)          'routine to flash an LED by color
    outa[color] :=high     'line that actually sets the LED high
    pause (200)            'wait till counter reaches this value
    outa[color] :=low      'line that actually sets the LED low
    pause (200)            'wait till counter reaches this value
'

PUB PAUSE(millisecs)               'As set up here it is 1.0 milliseconds
  waitcnt((clkfreq/1_000)*millisecs +cnt)      'ms based on Osc freq

PUB GetPotVal
  dira[PotLine]~~          'set potline as output
  valutotal:=0             'clear total
    repeat  repval         'repeat
      dira[PotLine]~~      'set potline as output
```

```
      outa[PotLine]~~        'make it high so we can charge the capacitor
      waitcnt(4000+cnt)      'wait for the capacitor to get charged
      dira[PotLine]~         'make potline an input. line switches H>L
      startCnt:=cnt          'read the counter at start of cycle and store
      repeat                 'go into an endless loop
      while ina[PotLine]~~   'keep doing it as long as the potline is high
      EndCount := cnt        'read the counter at end of cycle and store
      delay := ((EndCount-StartCnt)-1184)    'cal time for line to go H>L
      if delay>610_000         'max permitted delay
        delay:=610_000         'clamp delay
      PotValue:=(delay/2000)  'reduces the value to 0-255 or 1 byte
      valutotal:=valutotal+potvalue    'figures total
      potvalue:=valutotal/repval 'figure average
      potvalue <#=255
      potvalue #>=0
  result:=PotValue       'figure average

PUB GetPotVal2
  dira[PotLine2]~~             'set potline as output
  valutotal2:=0                'clear total
    repeat  repval2            'repeat
      dira[PotLine2]~~         'set potline as output
      outa[PotLine2]~~         'make it high so we can charge the capacitor
      waitcnt(4000+cnt)        'wait for the capacitor to get charged
      dira[PotLine2]~          'make potline an input. line switches H>L
      startCnt2:=cnt           'read the counter at start of cycle and store
      repeat                   'go into an endless loop
      while ina[PotLine2]~~    'keep doing it as long as the potline is high
      EndCount2 := cnt         'read the counter at end of cycle and store
      delay2:= ((EndCount2-StartCnt2)-1184)    'cal time for line to go H>L
      if delay2>610_000               'max permitted delay
        delay2:=610_000               'clamp delay
      PotValue2:=(delay2/2300)        'reduce the value to 0-255
      valutotal2:=valutotal2+potvalue2    'figures total
      potvalue2:=valutotal2/repval2 'figure average
      potvalue2 <#=255                'clamp at 255
      potvalue2 #>=0                  'clamp at 0
      result:=Potvalue2               'use average

PUB Read3202_0
  DIRA[chipSel]~~      'osc once to set up 3202
  DIRA[chipDin]~~      'data set up to the chip
  DIRA[chipDout]~      'data from the chip to the Propeller
  DIRA[chipClk]~~      'oscillates to read in data
    DataRed:=0              'Clear out old data
    outa[chipSel]~~          'Chip select has to be high to start off
    outa[chipSel]~           'Go low to start process
```

```
    outa[chipClk]~           'Clock needs to be low to load data
    outa[chipDin]~~          'must start with Din low to set up 3202
    outa[chipClk]~~          'Clock high to read data in

    outa[chipClk]~           'Low to load
    outa[chipDin]~~          'High single mode
    outa[chipClk]~~          'High to read

    outa[chipClk]~           'Low to load
    outa[chipDin]~           'low channel 0
    outa[chipClk]~~          'High to read

    outa[chipClk]~           'Low to load
    outa[chipDin]~~          'MSBF high = MSB first
    outa[chipClk]~~          'High to read

    outa[chipDin]~           'making line low for rest of cycle
    outa[chipClk]~           'Low to load Read the null bit, not stored
    outa[chipClk]~~          'High to read

    repeat BitsRead          'Reads the data into DataRed in 12 steps
      DataRed <<= 1 'Move data by shifting left 1 bit. Ready for next bit
      outa[chipClk]~         'Low to load
      DataRed:=DataRed+ina[chipDout]  'Xfer the data from pin chipDout
      outa[chipClk]~~        'High to read
    outa[chipSel]~~          'Put chip to sleep, for low power
    PotReading1:=DataRed     'Finished data read for display
    result:=dataRed
PUB Read3202_1
  DIRA[chipSel]~~      'osc once to set up 3202
  DIRA[chipDin]~~      'data set up to the chip
  DIRA[chipDout]~      'data from the chip to the Propeller
  DIRA[chipClk]~~      'oscillates to read in data
    DataRed:=0               'Clear out old data
    outa[chipSel]~~          'Chip select has to be high to start off
    outa[chipSel]~           'Go low to start process

    outa[chipClk]~           'Clock needs to be low to load data
    outa[chipDin]~~          'must start with Din low to set up 3202
    outa[chipClk]~~          'Clock high to read data in

    outa[chipClk]~           'Low to load
    outa[chipDin]~~          'High single mode
    outa[chipClk]~~          'High to read

    outa[chipClk]~           'Low to load
    outa[chipDin]~~          'high channel 1
    outa[chipClk]~~          'High to read
```

```
outa[chipClk]~           'Low to load
outa[chipDin]~~          'MSBF high = MSB first
outa[chipClk]~~          'High to read

outa[chipDin]~           'making line low for rest of cycle
outa[chipClk]~           'Low to load Read the null bit, not stored
outa[chipClk]~~          'High to read

repeat BitsRead2 'Reads the data into DataRed in 12 steps
  DataRed <<= 1  'Move data by shiftg left 1 bit. Rdy for nxt bit
  outa[chipClk]~          'Low to load
  DataRed:=DataRed+ina[chipDout]  'Xfer the data from pin chipDout
  outa[chipClk]~~         'High to read
outa[chipSel]~~           'Put chip to sleep, for low power
PotReading2:=DataRed      'Finished data read for display
result:=dataRed
```

B

MATERIALS

Table B-1 provides a list of all the hardware you will need to conduct all the experiments in this book, along with their approximate costs.

TABLE B-1 LIST OF HARDWARE ITEMS NEEDED

QTY	DESCRIPTION	SUPPLIER	APPROXIMATE COST
1	The Propeller Education Kit	Parallax	$99.99
4	7404 buffer gates	Jameco	$3.00
1	7805 voltage regulator	Jameco	$0.75
1	Motor amplifier: Xavien or Solarbotics	Encodergeek	$40.00
2	5K or 10K potentiometers. (Note: The kit comes with two small ones but we need larger ones.)	Jameco	$3.00
1	Small 12-VDC motor	Junk drawer	$2.00
1	Bipolar stepper motor	Jameco	$15.00
1	12-VDC motor with encoder	Encodergeek	$40.00
1	9-volt wall transformer	Junk drawer	$7.00
1	18-volt wall transformer	Junk drawer	$8.00
1	Gravity sensor: Memsic 2125	Parallax	$29.99
1	Small speaker	Misc	$2.00

C

TURNING COGS ON AND OFF

Although not exactly a beginner's subject, starting and stopping a cog may be a part of a program you are writing. Program C-1 can be used as a template for what you design.

This code is provided by Parallax (by SL) to demonstrate the starting and stopping of cogs using a few lines of code. Note that I have extended the code and documented each line to make it easier for the beginner to understand.

Program C-1 Starting and Stopping Cogs (Sample Code)

```
{{21 Dec 09      Harprit Sandhu
CogOnOff.spin
Propeller Tool Version 1.2.6
Appendix

Test Cog zero shutdown and restart.spin
Cog 0 starts blinking P4 LED at 3 Hz and Cog 1 starts blinking P5
LED at 10 Hz.  After 12.5 blinks, Cog 1 shuts down Cog 0.  After
50 blinks, Cog 1 launches LED blinking process into Cog 0 that blinks
the P6 LED at 23 Hz.
}}

VAR                                  'variables block
long stack[30]                       'stack space for second led

PUB Main                             'main cog
  Coginit (1, SecondLed, @stack)     'initializes and starts cog
  dira[4]~~                          'sets direction of line 4
  repeat                             'loop, infinite
    !outa[4]                         'make line 4 an output
    waitcnt(clkfreq/3 + cnt)         'wait for a third of second
```

(*continued*)

Program C-1 Starting and Stopping Cogs (Sample Code) (*continued*)

```
PUB SecondLed | counter                  'Cog for second led on line t
  counter~                               'set low = 0
  dira[5]~~                              'sets direction of line 4
  repeat                                 'loop
    !outa[5]                             'Toggle line 5
    waitcnt(clkfreq/10 + cnt)            'wait for a 1/10 of second
    counter++                            'increment counter
    if counter == 25                     'if counter is 25
      Cogstop(0)                         'Cog 0 stopped
    elseif counter == 50                 'test for 50
      Coginit(0, ThirdLed, @stack[15]) 'starts up cog 0 again
                                         '
PUB ThirdLed | counter                   'Cog for third led
  dira[6]~~                              'makes line an output
  repeat                                 'loop
    !outa[6]                             'toggles line
    waitcnt(clkfreq/23 + cnt)            'wait for a 1/23 of second
```

D

EXPERIMENTS BOARD

If you need a more robust platform or if you do not want to invest in the Education Kit (although I strongly recommend that you do), you can design a board that follows the circuitry shown in Figure D-1 and modify it to suit you needs. The circuit does not show the power supply wiring. Both populated and unpopulated single boards can be purchased from Encodergeek.com. You can populate the board over a period of time, as your expertise demands, to add the functionality you need. This approach will help you keep your initial costs down.

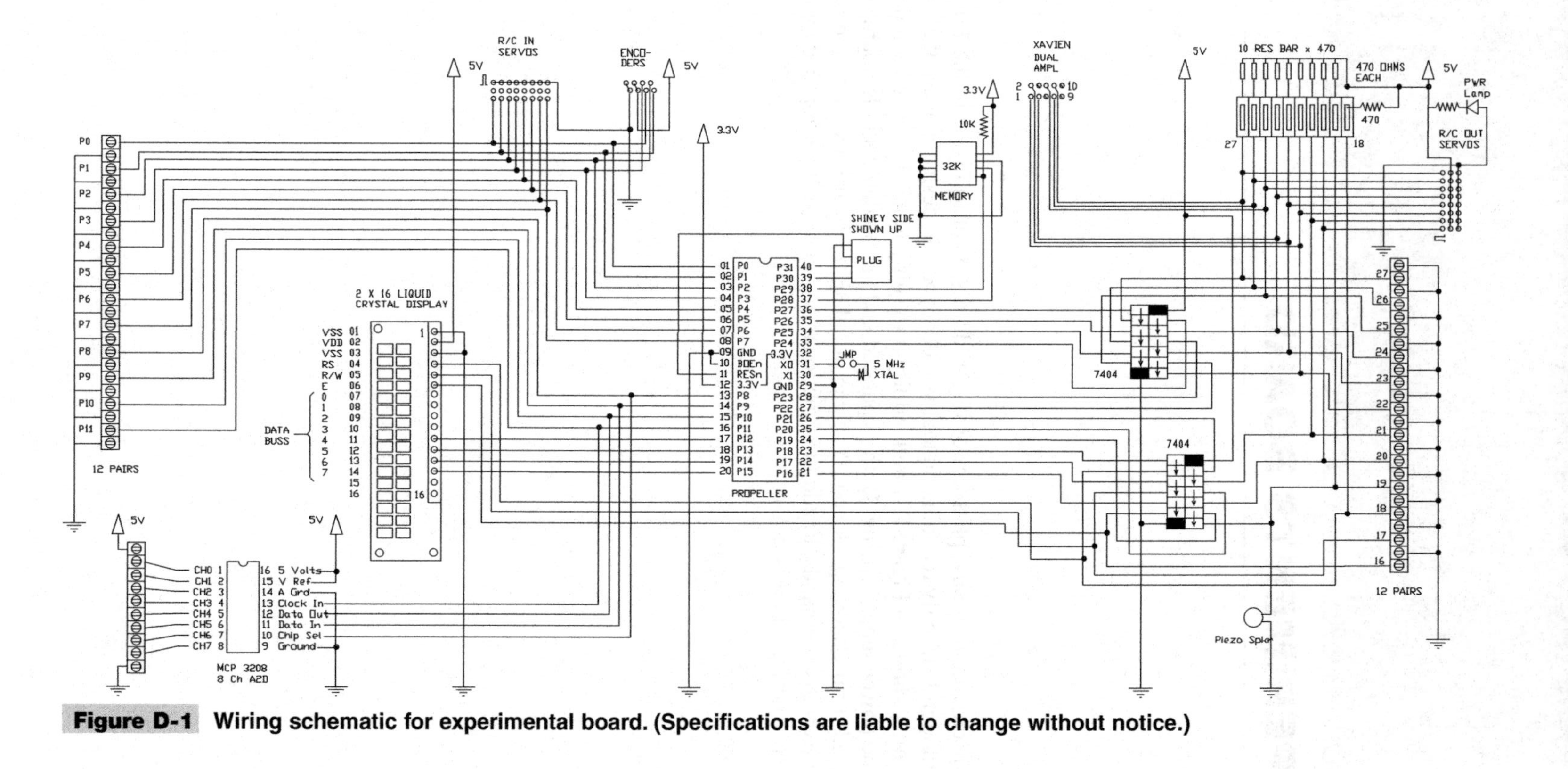

Figure D-1 Wiring schematic for experimental board. (Specifications are liable to change without notice.)

Figure D-2 Populated experimenter's board. (Prototype shown. Specifications are liable to change without notice.)

Figure D-2 shows the prototype. Improvements have been made in the production models. An unpopulated board (available from Encodergeek.com) can be populated as you need more features to keep your initial costs down.

E

DEBUGGING

Debugging and Troubleshooting

Debugging is not a random process during which one might hope to get lucky. It is a very carefully thought-out strategy to find why a program is not behaving the way it was intended to and what needs to be done to correct the problem. You will have fixed the problem only if you can make the problem come back by undoing the fix. A vague superstition that you might have fixed the problem by pressing on a warm resistor is not enough to decide that the problem has been fixed. You must be able to make the problem come back. This is exactly the reason why intermittent problems are so hard to fix. Its hard to make them come and go on command and so it is hard to understand the problem.

First Problem That Must Be Fixed *The microcontroller circuitry must oscillate.*

The clock frequency specification is important and must be specified accurately. The clock frequency specification for the Propeller is a little more complicated than it is for the average microprocessor. If you do not specify anything, it defaults to an internal frequency determined by a resistor capacitor pair inside the Propeller. No external components need to be attached to the Propeller in this case. The frequency will be approximately 4 MHz because the resister capacitor values in a mass-produced chip cannot be guaranteed to be exact.

If you are using a crystal across pins 25 and 26, you must specify the frequency exactly as suggested in the Propeller manual.

The oscillations can be checked with an oscilloscope. You should see a very clean sine wave (not square, if you are using a slower oscilloscope). Once you are satisfied you have proper oscillation, we can move to the other components.

Start off with the following checks on the hardware side:

1. Make sure the microcontroller has the correct power.
2. Make sure the power is 3.3 volts on the money.
3. Make sure you have clean DC power with no noise.
4. Use a magnifying glass to check the PC board for shorts and dry solder joints. Go over questionable areas with a soldering iron as you recheck your work.
5. Make sure the wiring is what you think it is. Check the route of every wire. Mark if off on the schematic as you check it. Check the PC board trace routings where necessary.
6. Check the values of each of the components on the board.
7. Make sure each IC is oriented with pin 1 in the proper location in its socket and that no pins are bent and thus not connected.
8. Make sure power and ground to each IC are properly routed.
9. Measure voltages throughout the layout and confirm that they are what they are supposed to be.
10. Make sure all capacitors are installed correctly. Check polarity where necessary.
11. Make sure all diodes are installed correctly. Not all always need to be installed so that the cathode is connected to ground. Confirm connections on all inductive loads.
12. Make sure all inductive loads are properly protected against with diodes. Make sure these diodes are properly rated for amperage, voltage, and switching speed. Make sure none have been destroyed.
13. Use the oscilloscope to check for noise all over the circuitry. Eliminate it by adding small capacitors near the noisy areas.
14. Read the relevant section of the manual (the startup section) again. Make sure you understand what the manual is saying.

Now make the following checks on the software side:

1. Go over the software, line by line, and make sure there are no typos. Not all typing mistakes may be identified by the software.
2. Write a short LED blink routine and run that on the board to make sure the system is actually alive and working. If the program loops as a part of its design, add a blinking instruction for one of the LEDs to tell you that the loop is actually executing as designed and not hanging up on some segment of code.
3. Go over the software to make sure there are no logical mistakes.
4. Follow the use of each variable throughout the program to make sure it does not get modified where it's not supposed to be modified.
5. Our software uses integer math. Make sure that none of the variables exceed the bounds that have been designated for their sizes.

There are a number of ways to get input into and *feedback out of from a malfunctioning program*. Using as many as possible will reduce the number of times you have to reload the program. You have the following feedback devices available to you. Incorporate what you need into critical areas of the program.

- **LCD** Use both lines on the LCD and use each character on each line.
- **Speaker** A speaker is easy to add. You can set up two tones that are easy to differentiate. Call them from various places in the program to see what is going on.
- **LEDs** If more is needed, add LEDs to the hardware so you can turn them on and off at critical junctures.

You can insert a *short loop that displays the registers you are interested in* at a critical location in the program. When the program enters this loop, it indicates that the program actually got this far and then it displays the registers of interest again and again without going any further. This loop can be moved up and down throughout the program so you can see what is going on and where. Here are the steps to follow:

1. Determine if the program is actually getting to a certain critical line of code.
2. Determine what the contents of various registers are at critical times in the code.
3. Look at how counters are behaving and confirm for yourself that this is exactly what is supposed to be happening.
4. Look for areas where the program might be getting stuck in a loop.
5. Pay special attention to the operation of various cogs and how they interact with variables specified by each other.
6. Go over the circuit layout to make sure there are no mistakes in the design of the circuitry.
7. Go over the physical circuits to make sure they are actually wired the way they were designed to be.
8. Make sure all lines that are to be pulled up or down are actually being pulled up or down and that the resistors being used are of the right values.

Dumb Terminal Program

A dumb terminal program can be used to send all sorts of data to the monitor as your program runs. The more you have to look at, the more likely it is that you will spot the problem.

A number of terminal programs are available at no charge on the Internet. I use the dumb terminal program provided by Microsoft as a part of its operating system utilities.

The Bray Terminal program is a more sophisticated dumb terminal program that's available for free on the Internet.

Signal Injection Techniques

When you're looking at various points on the circuit with an oscilloscope, mostly what you see are lines that are high or low. Most of the time that is not enough to recognize the presence or absence of a signal being traced. This can easily be remedied by creating a signal that provides a square wave that can be injected into the system. Looking at the original signal with the oscilloscope will tell you what to look for at various follow-up locations on the circuit board. Program E-1 provides a code segment that can be used in a spare cog to create a signal on pin P25.

Program E-1 Program to Generate a Frequency on One Line

```
CON
  _CLKMODE=XTAL1+ PLL2X             'The system clock spec
  _XINFREQ = 5_000_000D             '
  blink_pin     =25 ''''            '

PUB Go                              'main cog
  Cognew(cog_UpDn, @stack)           'start new cog for osc
      Repeat                           'parking loop

PRI wait                            'delay method
    waitCnt((clkfreq/12_000)*waitPeriod + cnt)   'specified by waitPeriod

PRI Cog_UpDn                        'osc cog
  dira[blink_pin]~~                 'make pin output
  repeat                            loop
    wait                            'wait for time
    !outa[blink_pin]                'invert line
```

Notes on Solderless Breadboards

Although we are using a system of solderless breadboards in our setup, the use of solderless boards for your prototyping activities is "in general" not recommended. They are alright for small experimental excursions when you first start out with the Propeller, but as your circuitry gets more and more sophisticated and complicated, there is too great a probability of poor connections and wires coming loose to use these devices.

It is recommended that you use the perforated boards that have a separate solder pad for each hole and then solder the wire wrap socket for each component into the board. Next, connect each of the components with wire wrap technology with straightforward point-to-point connections. The key is to be very careful and thorough so that there are no mistakes. It takes patience and care. Take your time and check your work before and after each connection is made. Mark it off on your circuit diagram one line at a time.

I also use circuit boards with continuous bars of conductors on them for some of my projects.

Debugging at the More Practical Level

Now what do I do? I've tried everything. The project is deader than a door nail and I don't have a clue!

A fairly long program you wrote will not work the way it is supposed to. You don't know if it is the software or the hardware, and you don't have a clue as to what you should do. Don't throw it all in the garbage just yet. Chances are that with a little bit of work, everything will work just as you intended. After all, you did create all the hardware and all the code, and hopefully no one knows more about it than you do!

The problem is that there is nothing to look at or to see: The thing is dead, and you don't know where to start. The solution is to make things visible and to start the process in a step-by-step manner so that you can make sure each step in the program you have created is doing what it is supposed to do. The good news is that you do not have to spend a fortune on new software and hardware, and you don't have to spend a year of your life learning a new discipline. You already know and have everything you need to debug the program in your Propeller.

The two output devices on the education board can be used as aids in the debugging process. These devices are:

- The LCD
- The LEDs placed on the board by you

Also, a number of input devices can aid in the debugging process by making the debugging process more interactive. These devices include:

- Potentiometers
- Buttons

We also have some of the standard software tools we can use, including the following:

- The Wait command
- The STOP command
- The IF…ELSE structure
- The ability to comment out sections of code

The PC you are using is also a powerful debugging tool in that it can both send and receive information and gives you a full screen and a keyboard to use as interactive elements. The object exchange maintained by Parallax on the Internet provides a number of powerful tools to let you interact with your PC. However, the most powerful device at your command is the computer between your ears. By and large, the debugging process is an exercise in the use of the brain, and everything else that needs to happen can be done with the Propeller/Spin system and your personal computer.

Here are some rules you need to follow to make the debugging process both easier and more likely to succeed within a reasonable amount of time:

- Rule 1: Be thinking about the debugging process as you write the code. Design the code so that it can be debugged. Be sure to put in the necessary hooks and connects as you go along.
- Rule 2: Write the code using small methods that can be tested as standalone mini-programs. Once you have the software working, you can streamline the code. Test your program as you develop it to make sure each developmental level is operational.
- Rule 3: Do not wait until the last moment to start the debugging process. Debug as you go along. In other words, you should debug the code as it is developed rather than waiting until it is all done and ready to be delivered to the customer. Learn to write your code so that you can debug sections of it as sections of code or as standalone methods.
- Rule 4: Write a set of routines that can be called from within the code that shows you the content of various memory locations on the LCD as the program runs.

The first thing most programs must do is make the LCD come alive. Two questions have to be answered to confirm its proper operation:

- Is the software right?
- Have you somehow destroyed the electronics?

Writing a Rudimentary Program for Testing the LCD

A comprehensive program should be in your utility files to allow you to check the proper hardware and software operation of your LCD whenever you think it is necessary. Your program should check every character and every command in the LCD's vocabulary if you want to perform a really comprehensive check. You may even want to check every pixel.

Integer math is the source of a lot of problems (for those who are unaware of the havoc that integer math calculations can visit upon the software being debugged). A certain amount of expertise with integer math is a must if you are going to create mathematical routines within your software. If the routine is amenable to it, you should write a program around the routine to test every possibility that the routine might encounter and thus debug it the hard way, even if it means your computer has to run the routine all night to get through all the commutations. Oftentimes all that is necessary is to run the routines that would be called at the boundary conditions or under critical conditions to make sure they are robust.

Register underflows and overflows are a common problem. You have to make sure they do not occur. The system usually just crashes and does not provide any guidance as to what the problem might be when an underflow or overflow happens.

Another List of Simple Checks

Here are some simple checks you should apply if the Propeller still does not oscillate:

- ❑ Check that the power is on to the project.
- ❑ Check that the USB plug is connected to the computer.
- ❑ Check that the USB plug is plugged into your board.
- ❑ Check that the Propeller is properly orientated for pin 1.
- ❑ Check that the program has proper code indentation (very important).
- ❑ Check that the addresses for all the devices for the project are correct.
- ❑ Check the spelling of all the code.
- ❑ Check that all sockets that should be empty are actually empty.
- ❑ Run a test program suited to and designed for your specific project.
- ❑ Check with an oscilloscope that the system is actually oscillating at the OSC pins.
- ❑ In most programs, some other pins should also be going high and low regularly.
- ❑ Recheck the wiring.
- ❑ Check for open wires and cold solder joints.
- ❑ Check for short circuits.
- ❑ Recheck the program.
- ❑ Run a program you know works on a Propeller you know works.

Good luck helps, but there is nothing that beats having done your homework.

EPILOGUE

This book tries to do as much as possible with just the Propeller and the education kit Parallax provides for it. It minimizes the use of external hardware as much as possible, and it minimizes the number of Spin language commands used. I wrote the book this way to make it as easy as possible for beginners to get comfortable with using the Propeller chip. Once you understand the basics, you should have no problem with adding hardware and software instructions to your projects.

The education kit can be reused with the very good, although a bit more advanced, teaching material Parallax provides. A lot of the information you will need can be downloaded from the Internet at no charge, so the education kit can be used over and over again. I doubt that one could find and purchase all the odds and ends that come with the education kit for less than the cost of that kit.

I feel that learning by doing is the best way to learn, so I tried to provide as much code as I could for each of the experiments to get you comfortable with reading programs. The code is broken into smaller segments, where appropriate, to make it easier to understand. Almost every line of code is commented—some with more than one line—to indicate exactly what each line of code is doing. Hopefully, this will encourage you to extensively comment the code you write so that both you and others who use your code will be comfortable with it and can understand what is being accomplished. You will be surprised by how much you forget about why you wrote what you did after a couple months have passed. Good, well-thought-out, extensive documentation is very important to the success of your efforts. Much of the code available on the Internet is not documented at all, and a lot more has but minimal documentation. I am a beginner as far as the Propeller system goes, and I found it rather difficult to understand what was going on in many of the programs I studied.

I freely admit that the Assembly language programs I tried to understand were completely "Greek" to me. Even in Spin programs, where shorthand notation was used or where more than one instruction was combined on one line, I found myself at a loss. Therefore, I tried to make sure that this practice was *not* repeated in the book. This tends to make the programs look a bit simple minded, but hopefully they can be understood without you pulling your hair out. Once you understand what is going on, you can make your code as sophisticated and as complicated as you like.

—Harprit Singh Sandhu
Champaign, Illinois USA
May 2010
harprit.sandhu@gmail.com

INDEX

D

E

L

M

N

O

Q

R

S

T

U

V

W

X

CPSIA information can be obtained
at www.ICGtesting.com
Printed in the USA
LVOW04s1724210717
542007LV00006B/43/P